CAMBRIDGE LIBRARY COLLECTION

Books of enduring scholarly value

Polar Exploration

This series includes accounts, by eye-witnesses and contemporaries, of early expeditions to the Arctic and the Antarctic. Huge resources were invested in such endeavours, particularly the search for the North-West Passage, which, if successful, promised enormous strategic and commercial rewards. Cartographers and scientists travelled with many of the expeditions, and their work made important contributions to earth sciences, climatology, botany and zoology. They also brought back anthropological information about the indigenous peoples of the Arctic region and the southern fringes of the American continent. The series further includes dramatic and poignant accounts of the harsh realities of working in extreme conditions and utter isolation in bygone centuries.

The Zoology of the Voyage of H.M.S. *Herald*

Naval surgeon, Arctic explorer and natural historian, Sir John Richardson (1787–1865) published many works, several of which are reissued in the Cambridge Library Collection, notably the four-volume *Fauna Boreali-Americana*. At the Haslar Royal Naval Hospital, where he worked towards the end of his career, Richardson built up a library and museum that became renowned for natural history research. His published work was fuelled by his own voyages and the specimens sent back from other expeditions, as was the case for this illustrated work, completed in 1854. Richardson describes the zoological specimens collected during the 1845–51 voyage of the survey ship H.M.S. *Herald*, which had sailed into Arctic seas and took part in the search for Sir John Franklin. The collected fauna include fossil mammals from the ice cliffs at Eschscholtz Bay in Alaska, first discovered in 1816 by Otto von Kotzebue and his naturalists.

Cambridge University Press has long been a pioneer in the reissuing of out-of-print titles from its own backlist, producing digital reprints of books that are still sought after by scholars and students but could not be reprinted economically using traditional technology. The Cambridge Library Collection extends this activity to a wider range of books which are still of importance to researchers and professionals, either for the source material they contain, or as landmarks in the history of their academic discipline.

Drawing from the world-renowned collections in the Cambridge University Library and other partner libraries, and guided by the advice of experts in each subject area, Cambridge University Press is using state-of-the-art scanning machines in its own Printing House to capture the content of each book selected for inclusion. The files are processed to give a consistently clear, crisp image, and the books finished to the high quality standard for which the Press is recognised around the world. The latest print-on-demand technology ensures that the books will remain available indefinitely, and that orders for single or multiple copies can quickly be supplied.

The Cambridge Library Collection brings back to life books of enduring scholarly value (including out-of-copyright works originally issued by other publishers) across a wide range of disciplines in the humanities and social sciences and in science and technology.

The Zoology of the Voyage of H.M.S. *Herald*,

Under the Command of Captain Henry Kellet, During the Years 1845–51

Vertebrals, including Fossil Mammals

JOHN RICHARDSON
EDITED BY EDWARD FORBES

CAMBRIDGE
UNIVERSITY PRESS

CAMBRIDGE
UNIVERSITY PRESS

University Printing House, Cambridge, CB2 8BS, United Kingdom

Cambridge University Press is part of the University of Cambridge.

It furthers the University's mission by disseminating knowledge in the pursuit of
education, learning and research at the highest international levels of excellence.

www.cambridge.org
Information on this title: www.cambridge.org/9781108073684

© in this compilation Cambridge University Press 2014

This edition first published 1852–4
This digitally printed version 2014

ISBN 978-1-108-07368-4 Paperback

FOSSIL MAMMALS.

BY

SIR JOHN RICHARDSON, Knt. C.B., M.D., F.R.S.

THE ZOOLOGY

OF THE

VOYAGE OF H.M.S. HERALD,

UNDER THE COMMAND OF

CAPTAIN HENRY KELLETT, R.N., C.B.,

DURING THE YEARS 1845-51.

Published under the Authority of the Lords Commissioners of the Admiralty.

EDITED BY

PROFESSOR EDWARD FORBES, F.R.S.

FOSSIL MAMMALS.

BY

SIR JOHN RICHARDSON, Knt. C.B., M.D., F.R.S.

LONDON:
REEVE AND CO., 5, HENRIETTA STREET, COVENT GARDEN.

1852.

PRINTED BY
JOHN EDWARD TAYLOR, LITTLE QUEEN STREET,
LINCOLN'S INN FIELDS.

CONTENTS

OF PARTS I., II., III.,

EMBRACING FOSSIL MAMMALS, AND RECENT REPTILES AND FISH.

———◆———

MAMMALS.

	PAGE.
On the Structure of the Fossil Bone Deposit in Eschscholtz Bay	1–8
Notice of Technical Terms employed in Comparative Anatomy by Dr. Barclay	9
Fossil remains of the Mammoth (*Elephas primigenius?*) from Eschscholtz Bay	11–17
Fossil remains of the Mammoth from Yukon River, lat. 61° N.	142
Fossil remains of the *Mastodon giganteus* from Swan River, lat. 49° (erroneously *Elephas Rupertianus*)	101–102, 141–142
Fossil remains of the Horse (*Equus fossilis*)	17–20
Fossil remains of the Moose-deer (*Alces*)	20
Osteology of recent Moose-deer (*Alces Muswa**)	102–114
Fossil remains of the Rein-deer (*Cervus tarandus*) from Eschscholtz Bay	20–21
Osteology of recent Rein-deer, or "Tuktu," from the Arctic Circle	98, 99, 115–119
Fossil remains of the Musk-ox (*Ovibos moschatus*) from Eschscholtz Bay	22, 25
Osteology of recent Musk-oxen	66–87, 119
Fossil *vertebra dentata* of *Ovibos maximus*	25–28
Its affinity to *Bootherium cavifrons* of Leidy surmised	139
General remarks on the *Bovidæ* (compiled)	28–32
General remarks from the papers of Dr. Leidy, on *Bison Americanus, B. latifrons, B. antiquus, Bootherium bombifrons*, and *Booth. cavifrons*	139
Fossil remains of the restricted genus *Bos* not yet found in America	139
On the fossil remains of a Bison from Eschscholtz Bay (*Bison priscus?*), proper specific appellation uncertain	33–40
On the fossil remains of *Bison crassicornis*, or the "Heavy-horned fossil Bison"	40–61
Its affinity to or identity with *Bison antiquus* of Leidy surmised	139
On a mutilated fossil cervical, No. 118, more nearly resembling that of the existing "Mustush," or *Bison Americanus*, than that of any other known species, but scarcely identical with it	61
Osteology of the Aurochs (*Bison Europæus*): additional remarks on some joints	99
Osteology of the Aurochs (*Bison Europæus*): further measurements of the skeleton	122

* *Alces machlis* is Dr. Gray's appellation for the European Elk, the specific name being that used by Pliny.

vi   CONTENTS.

PAGE.

Some remarks on the state of the fossil bones in the Ice-cliffs of Eschscholtz Bay . . . 62
Osteology of a Big-horn Ram (*Ovis montana*) aged twenty-two months . . . 87–98
Osteology of the Rocky Mountain Antelope (*Aplocerus montanus*) 131–138

REPTILES.

Lophosaura Goodridgii (Gray) 143–146
Osteology of *Lophosaura Goodridgii* 146–148
Craneosaura Seemanni (Gray) 148–150
Osteology of *Craneosaura Seemanni* 150–151
Gecko Reevesii (Gray) 151–152
Osteology of *Gecko* (*verus?*) 152–154
Anniella pulchra (Gray) 154–155
List of Reptiles and Batrachians collected on the voyage 155–156

FISH.

Anchisomus geometricus (Kaup) 156–157
Osteology of *Anchisomus geometricus* 157–158
Anchisomus angusticeps (Jenyns) 159–160
Anchisomus multistriatus (Kaup) 160–161
Anchisomus reticularis (Kaup) 161–162
Prilonotus (*an Anchisomus?*) *caudacinctus* 162–163
Tetrodon virgatus 163–164
Platessa stellata (Pallas) 164–165
Platessa glacialis (Pallas) 166–167
Salmo consuetus (Richardson) 167–169
Salmo dermatinus (Richardson) 169–171

LIST OF PLATES.

———

(None of the figures were reversed when drawn on the stone, so that where the two sides differ their denominations of right or left are reversed.)

PLATE I.

Fig. 1. Skeleton of the Big-horn (*Ovis montana*) :—*one-fifth natural size.* (Pages 87–98.)

Fig. 2. Front view of the skull :—*one-fifth natural size.*

Fig. 3. Hind view of skull :—*one-fifth nat. size.*

PLATE II.

Fig. 1. Skeleton of a Musk-bull (*Ovibos moschatus*), four or five years old :—*one-fifth nat. size.* (Page 66 *et infra.*)

Fig. 2. Antinio-coronal view of the skull :—*one-fifth nat. size.*

PLATE III.

Inial view of the same Bull :—*nat. size.* (Page 68.)

PLATE IV.

Fig. 1. Coronal aspect of an adult, but not aged, Musk-cow :—*half nat. size.* (Page 67.)

Fig. 2. Coronal aspect of Musk-bull calf, thirteen months old, offspring of the preceding cow :—*half nat. size.* (Page 67.)

Fig. 3. Coronal aspect of a fœtal Musk-calf taken from the above-mentioned cow, supposed to be six months after conception : —*nat. size.* (Page 71.)

Fig. 4. Inial aspect of the same skull :—*nat. size.*

PLATE V.

Fig. 1. Sternal aspect of the *atlas* and *dentata* of the Musk-bull figured in Plate II. :—*nat. size.* (Page 73.)

Fig. 2. Dorsal (or neural) aspect of the same *atlas* :—*nat. size.*

Fig. 3. Proximal aspect of same *atlas* :—*nat. size.*

Fig. 4. Distal (or sacral) aspect of same *atlas* :—*nat. size.*

Fig. 5. Lateral aspect of the *dentata*, which is shown in Fig. 1 in conjunction with its *atlas* :—*nat. size.* (Page 73.)

PLATE VI.

Fig. 1. Inial aspect of the skull of the adult male Lithuanian Aurochs (*Bison Europœus*), part of the entire skeleton preserved in the British Museum :—*nat. size.* (Pages 33–42, *passim.*)

Fig. 2. Basilar aspect of the condyles and basi-occipital of the same skull :—*nat. size.*

Fig. 3. Inial aspect of the skull of an adult Mustush-bull (*Bison Americanus*), preserved in the British Museum :—*nat. size.* (Pages 34–42, *passim.*)

Fig. 4. Basilar aspect of the condyles and basi-occipital of the same Mustush skull :—*nat. size.*

Fig. 5. Basilar aspect of the condyles and basi-occipital of the fossil No. 24,576 (*Bison priscus ? ?*), found in the ice-cliffs of Eschscholtz Bay, and now in the British Museum :—*nat. size.* (Page 36.)

Fig. 6. Inial view of the condyles of the same fossil fragment :—*nat. size.*

PLATE VII.

Comparative views of the horn-cores and forehead

of a fossil skull from Eschscholtz Bay, with
those of a Lithuanian bull Aurochs and Mustush-bull, all reduced to *half their linear dimensions*.

Fig. 1. Fossil skull, No. 24,589 (*Bison priscus ? ?*),
British Museum :—*half nat. size*.

Fig. 2. Adult bull Aurochs (figured in Plate VI.
fig. 1) :—*half nat. size*. (Pages 33–43, *passim*,
and 64.)

Fig. 3. Lateral aspect of the same Aurochs skull :
—*half nat. size*.

Fig. 4. Skull of the adult Mustush (*Bison Americanus*), also figured in Plate VI. Fig. 3 :—*half
nat. size*. (Page 35, and 33–43, *passim*.)

Fig. 5. Lateral aspect of the same Mustush skull :
—*half nat. size*.

Plate VIII.

Views of the *atlas* and *dentata* of an adult Aurochs
bull and Mustush cow, both in the collection
of the British Museum. All reduced to *half
the linear size*.

Fig. 1. *Atlas* of the *Aurochs* bull, sternal aspect.
(Pages 43, 37, and 73.)

Fig. 2. *Atlas* of the *Aurochs* bull, lateral aspect.

Fig. 3. *Atlas* of the *Aurochs* bull, oblique view of
the proximal and sternal aspects.

Fig. 4. *Atlas* of the *Aurochs* bull, sacral aspect.

Fig. 5. *Atlas* of a young *Mustush* cow, sternal aspect. (Pages 43, 44.)

Fig. 6. *Atlas* of a young *Mustush* cow, lateral
aspect.

Fig. 7. *Atlas* of a young *Mustush* cow, oblique view
of the proximal and sternal aspects.

Fig. 8. *Atlas* of a young *Mustush* cow, sacral
aspect.

Fig. 9. *Dentata* of the *Aurochs* bull, sternal aspect.
(Pages 65, 123, 124.)

Fig. 10. *Dentata* of the *Aurochs* bull, lateral aspect.

Fig. 11. *Dentata* of the *Mustush* cow, sternal aspect.

Fig. 12. *Dentata* of the *Mustush* cow, lateral aspect.

Plate IX.

Views of portions of the skull of *Bison crassicornis*, a fossil species mainly founded on the
large horn-cores of Plate XIII. Fig. 1 and 2,

and on the skull here represented, now in the
British Museum.

Fig. 1. Forehead and horn-cores, drawn from the
same point of view with the other species
figured in Plate VIII. :—*half nat. size*. (Pages
40–42.)

Fig. 2. Inial view of the same skull (the parts
within the dotted line, left unfinished by the
artist, whose work I was unable to super-
intend) :—*nat. size*. (Page 42.)

Fig. 3. Condyles and basi-occipital of the same
skull :—*nat. size*.

Fig. 4. Lateral aspect of same skull :—*nat. size*.

Plate X.

Figures of bones belonging to the fossil Bison of
the ice-cliffs of Eschscholtz Bay, named in the
text *Bison priscus*, but which is seemingly a
peculiar species, perhaps confined to America ;
its bones differing from those of the Lithuanian
aurochs, as well as from the American bison or
mustush, but most resembling the latter.

Fig. 1. Lateral view of the hinder part of a *skull*,
forming part of Captain Kellett's first collection, No. 24,589 Br. Mus. Cat. of fossils (a
front view of the forehead is given in Plate
VII. fig. 1) :—*half nat. size*. (Pages 33–35.)

Fig. 2. Sternal view of an *atlas* from the same locality, No. 24,576, Br. Mus. Cat. of fossils :—
nat. size. (Page 37.)

Fig. 3. Dorsal aspect of the same :—*nat. size*.

Fig. 4. Proximal aspect of the same :—*nat. size*.

Fig. 5. Distal aspect of the same :—*nat. size*.

Fig. 6. Oblique view of the interior of the neural
canal, to show the absence of a groove between
the lateral pits of the centrum :—*nat. size*.

Plate XI.

Fossil and recent vertebræ, all of the *nat. size*.

Fig. 1. *Ovibos moschatus*, recent. Sacral view of
the *dentata* of a four-to-five-year-old Musk-
bull.

Fig. 2, 2. *Ovibos maximus*, fossil. Sacral view of a
dentata, No. $\frac{o\cdot o}{\frac{9}{2}}$, now in the Haslar Museum,
found in the ice-cliffs of Eschscholtz Bay by
Captain Kellett's party, and the solitary frag-

ment of the skeleton on which the species is founded. (Pages 25–28.)

Fig. 3. Sternal aspect of the same.

Fig. 4. Lateral aspect of the same. Neural spine and diapophyses broken away. By a slight error in the outline of the proximal articular surface, the mesial notch is thrown to the right of its true position.

Fig. 5. *Ovibos moschatus*, recent. Sacral view of the first dorsal of the four-to-five-year-old Musk-bull.

Fig. 6. *Bison crassicornis?* fossil. Sacral view of the first dorsal, No. 121 Haslar Museum. (Page 151.)

Plate XII.

Fig. 1. *Bison crassicornis.* Fossil *atlas* from Eschscholtz Bay, referred provisionally to this species: sternal aspect :—*nat. size.* No. 90 Haslar Museum. (Pages 43, 44, 45.)

Fig. 2. Dorsal aspect of the same :—*nat. size.*

Fig. 3. Proximal aspect of the same :—*nat. size.*

Fig. 4. Sacral aspect of the same :—*nat. size.*

Fig. 5. *Ovis montana.* Recent *atlas* of a Big-horn ram: oblique sternal view :—*nat. size.*

Fig. 6. Oblique dorso-sacral aspect of the same :— *nat. size.*

Plate XIII.

Fossil horn-cores from Eschscholtz Bay :—*nat. size.*

Fig. 1. *Bison crassicornis.* Antinial aspect. No. 91. (Pages 42, 43.)

Fig. 2. Coronal aspect of the same.

Fig. 3. Smaller fossil Bison of the Ice-cliffs, named *Bison priscus??* in the text; the core extracted from its horny sheath by boiling. No. 105 Haslar Museum. (Page 35.)

Plate XIV.

Recent Musk-cow (*Ovibos moschatus, fem.*). Lumbars of the same individual, whose skull is represented on Plate IV. Figures of *nat. size.* (Page 80.)

Fig. 1. Last four lumbars, dorsal aspect. The lowest in the Plate articulates with the sacral. In the penultimate one the epiphysial points of the pleurapophyses have separated

by maceration, but in the others the bony confluence was more advanced.

Fig. 2. Lateral view of the last three of the same series, the one to the right, next the numeral 2, being that which articulates with the sacral.

Figs. 3, 4, 5. Various views of the fourth lumbar of the same animal, being the third from the sacrum. The metapophyses are well seen in fig. 4. This vertebra is the second in fig. 1, and the first to the right in fig. 2 ; its dimensions are given on page 80.

Plate XV.

Recent and fossil bones, figured of *nat. size.*

Fig. 1. *Bison crassicornis?* Antepenultimate lumbar, atlantal aspect. No. 123 Haslar Museum. (Page 53.)

Fig. 2. Lateral aspect of the same.

Fig. 3. *Bison crassicornis?* Nos. $\frac{72}{1}$, $\frac{72}{2}$, Haslar Museum ; found with the preceding in the Ice-cliffs. *Radius* and *ulna;* the latter defective in the end of the olecranon and part of its shaft. Thenal aspect. (Pages 56–58.)

Fig. 4. Inside of the elbow-joint of the same fossil.

Fig. 5. *Ovibos moschatus.* *Radio-ulnar* joint of a recent Musk-cow, the same as that whose skull is represented in Plate IV. (Page 57.)

Plate XVI.

Lateral view of the skull and mandible of the Rocky Mountain Antelope, *Rupicapra* (vel *Aplocerus*) *Americana*, recent :—*natural size.* (Page 131.)

Plate XVII.

Different views of the subject of the preceding Plate :—*nat. size.* (Page 131.)

Fig. 1. Antinio-coronal aspect.

Fig. 2. Inial aspect.

Fig. 3. Basilar aspect.

Plate XVIII.

Cervicals of the same Antelope :—*nat. size.* (Pages 133, 134.)

Fig. 1. Lateral aspect.

Fig. 2. Sternal aspect.

PLATE XIX.

Some bones of the same skeleton :—*natural size.* (Pages 136–138.)

Fig. 1. Thenal aspect of the *radius* and *ulna.*

Fig. 2. Section of the same in the middle of the shafts.

Fig. 3. Anconal aspect of the atlantal cannon-bone, or (united third-fourth) metacarpal, with a rudimentary fifth metacarpal *in situ.*

Fig. 4. Section of the same in the middle of the shaft.

Fig. 5. Popliteal aspect of the (united third-fourth) metacarpals, or sacral cannon-bone. A circular, flat sesamoid, which lay in the shaded depression at the proximal end of the bone, has been removed.

Fig. 6. Section of the same in the middle of the shaft.

PLATE XX.

Alces Muswa, or American Moose-deer or Elk :— *nat. size.* (Page 102.) *Atlas, dentata,* and third cervical : sternal aspect.

PLATE XXI.

Lateral aspect of the same three cervicals :—*nat. size.* (Page 102.)

PLATE XXII.

Fig. 1. *Alces Muswa.* Inial view of the skull of a male with its antlers :—*one-fourth nat. size.* (Page 103.)

Fig. 2. *Rangifer,* "Tuktu," or Barren-ground Rein-deer : recent. Lateral view of the confluent third-fourth metacarpals, or atlantal cannon-bone, with the second and fifth metacarpals and their phalanges :—*nat. size.* (Page 106.)

PLATE XXIII.

Rangifer, "Tuktu," or Barren-ground Rein-deer : Antinio-coronal view of the skull of a male :— *one-fourth nat. size.* (Pages 98, 117.)

PLATE XXIV.

Alces Muswa : American Moose-deer, male :—*nat. size.* (Pages 111–114.)

Fig. 1. Glenoid cavity of the *scapula* with its coracoid and spinous processes.

Fig. 2, 2. Thenal aspect of the *radius* and *olecranon,* showing the elbow-joint. A section of the middle of the shaft is appended to the head of the second segment.

Fig. 3. Thenal aspect of the atlantal cannon-bone, or confluent third-fourth *metacarpals,* with the second and fifth metacarpals and their phalanges.

Fig. 4. Section near the middle of the cannon-bone.

PLATE XXV.

Craneosaura (perperam Ptenosaura) Seemanni (Gray). (Page 148.)

Fig. 1. Profile :—*nat. size.* (Page 148.)

Fig. 2. Coronal aspect of the head :—*natural size.* (Page 149.)

Fig. 3. Basilar aspect of the head :—*natural size.* (Page 149.)

PLATE XXVI.

Lophosaura Goodridgii (Gray).

Fig. 1. Profile :—*nat. size.* (Page 143.)

Fig. 2. Coronal aspect of the head :—*natural size.* (Page 144.)

Fig. 3. Basilar aspect of the head :—*natural size.* (Page 144.)

PLATE XXVII.

Gecko Reevesii (Gray).

Fig. 1. Oblique dorsal view :—*nat. size.* (Pages 151, 152.)

Fig. 2. Profile of the head :—*nat. size.* (Page 151.)

Fig. 3. Coronal aspect of the head :—*nat. size.* (Page 151.)

Fig. 4. Basilar aspect of the head :—*nat. size.* (Page 151.)

Fig. 5. Thenal aspect of the fore foot :—*nat. size.* (Page 152.)

Fig. 6. Plantar aspect of the hind foot :—*nat. size.* (Page 152.)

Fig. 7. Rotular aspect of the hind foot :—*magn.* (Page 152.)

Fig. 8. A toe of the hind foot : plantar or popliteal aspect :—*magn.* (Page 152.)

PLATE XXVIII.

Fig. 1. Profile of *Anniella pulchra* (Gray) :—*nat. size.* (Page 154.)

Fig. 2. Fore part of *Anniella pulchra* :—*magn.* (Page 154.)

Fig. 3. Coronal aspect of head and neck of *Anniella pulchra* :—*magn.* (Page 154.)

Fig. 4. Basilar aspect of the head of *Anniella pulchra* :—*magn.* (Page 154.)

Fig. 5. Anal region of *Anniella pulchra* :—*magn.* (Page 154.)

Fig. 6. *Tetrodon virgatus*, profile :—*natural size.* (Page 163.)

Fig. 7. Dorsal aspect of *Tetrodon virgatus* :—*nat. size.* (Page 164.)

Fig. 8. Ventral spines, and their containing sacs :—*magn.* (Page 164.)

PLATE XXIX.

Anchisomus multistriatus (Kaup) :—all the figures of *half tho nat. size.*

Fig. 1. Profile of the dried skin. (Page 160.)

Fig. 2. Dorsal aspect of fore part of the dried skin. (Page 160.)

Fig. 3. Two of the spines.

PLATE XXX.

Anchisomus geometricus (Kaup).

Fig. 1. Profile :—*nat. size.* (Page 156.)

Fig. 2. Dorsal aspect, without the spots to show the spines :—*nat. size.* (Page 156.)

Fig. 3. Nuchal spines :—*nat. size.* (Page 156.)

Fig. 4. Ventral spines :—*nat. size.* (Page 156.)

Fig. 5. Ventral spines :—*magn.* (Page 156.)

PLATE XXXI.

Fig. 1. *Prilonotus* (vel *Anchisomus ?*) *caudacinctus*, profile :—*nat. size.* (Page 162.)

Fig. 2. Dorsal aspect of *Prilonotus* (vel *Anchisomus?*) *caudacinctus*, faintly rendered to show the spines :—*nat. size.* (Page 162.)

Fig. 3. Ventral spines of *Prilonotus* (vel *Anchisomus?*) *caudacinctus*, showing through the integument, with the lacunæ from whence their points protrude :—*magn.* (Page 162.)

Fig. 4. *Anchisomus reticularis* (Kaup), profile :—*nat. size.* (Page 161.)

Fig. 5. Outline of dorsal aspect to show the spines :—*nat. size.* (Page 161.)

Fig. 6. Spines :—*magn.* (Page 161.)

PLATE XXXII.

Fig. 1. *Platessa stellata* (Pallas), view of the left side: *nat. size.* (Page 164.)

Fig. 2. View of the right side :—*nat. size.* (Page 164.)

Fig. 3. Scales, from bases of rays :—*magn.* (Page 164.)

Fig. 4. *Platessa glacialis* (Pallas), view of right side of the head :—*nat. size.* (Page 166.)

These figures were not reversed on the stone, and tho right sides are therefore represented as the left, and *vice versâ*, in the Plate.

PLATE XXXIII.

Fig. 1. *Salmo consuetus*, profile of the dried skin :—*half the nat. size.* (Page 167.)

Fig. 2. Tips of the jaws :—*half the natural size.* (Page 168.)

Fig. 3. *Salmo dermatinus*, profile of the dried skin :—*half the nat. size.* (Page 169.)

Fig. 4. Tips of the jaws :—*half the natural size.* (Page 170.)

Fig. 5. Cluster of scales from the side :—*magn.* (Page 170.)

ZOOLOGY

OF THE

VOYAGE OF H.M.S. HERALD.

———◆———

OBSERVATIONS ON THE FOSSIL BONE DEPOSIT IN
ESCHSCHOLTZ BAY.

THE science of chemistry, as at present taught, justifies our belief that animal substances, when solidly frozen and kept steadily in a temperature below the freezing-point, do not undergo putrefaction, and may be preserved without change for any conceivable length of time. The depth to which, in northern countries, the summer thaw penetrates, varies with the nature of the soil, but, except in purely sandy and very porous beds, it nowhere exceeds two feet in American or Siberian lands lying within the Arctic Circle. The influence of the sun's rays is not perceptible at this depth until towards the close of summer, which occurs at a varying period of from five to ten weeks from the time that the surface of the earth was denuded of snow by the spring thaw. During the rest of the year, even in the forest lands, though not so long there as in the open barren grounds, or *tundras*, the soil is firmly and continuously bound up in frost. The thickness of the permanently frozen substratum is more or less influenced by its mineral structure, but is primarily dependent on the mean annual temperature of the air acting antagonistically to the interior heat of the earth. Unless the mean heat of the year in any given locality falls short of the freezing-point, there exists no perennial frozen substratum at that place. It is not necessary that we should here endeavour to trace the isothermal line of 32° Fahr., as the reader may obtain a correct idea of its general course by consulting Baer's charts. It will suffice to say, that on the continent of America it passes some degrees to the southward of the sixtieth parallel of north latitude, and that while it undulates with the varying elevation of the interior, it has a general rise northwards in its course westerly.

Where the permanently frozen subsoil exists, it is a perfect ice-cellar, and preserves from destruction the bodies of animals completely enclosed in it. By its intervention entire carcases of the extinct mammoth and tichorhine rhinoceros have been handed down in arctic Siberia from the drift period to our times, and, being exposed by landslips, have revealed most interesting glimpses of the fauna of that remote epoch. Conjecture fails in assigning a chronological date to the time when the drift and boulders were spread extensively over the northern hemisphere : the calculations that have been made of the ages occupied in the formation of subsequent alluvial deposits are founded on imperfect data; and we merely judge from the absence of works of art and of human bones, that the drift era must have been antecedent to the appearance of man upon earth, or at least to his multiplication within the geographical limits of the drift. Whatever may be our speculations concerning the mode in which the carcases in question were enclosed in frozen gravel or mud, their preservation to present times in a fresh condition indicates that the climate was a rigorous one at the epoch of their entombment and has continued so ever since. Moreover, as large carcases could not, without decomposition, be conveyed from a distance by water, it is fair to conclude that the animals lived in the districts in which they are now found, or in their immediate neighbourhood, and not, as some have supposed, in warmer and more distant regions.

It seems also to us to be impossible that ice could have been the vehicle by which whole bodies or complete skeletons could have been brought from warmer parallels and deposited in the vast cemeteries of polar Siberia or in Eschscholtz Bay, for the simple reason that ice is not the product of these warm countries. Nor does the difficulty seem less of explaining how such a group of pachyderms and ruminants could have been brought down by travelling glaciers from warmer southern valleys of mountain ranges no longer in existence, without admitting such extensive changes in the surface level of the district, as would confound all our ideas of the distribution of the drift as we at present find it.

It is easier to imagine that the animals whose osseous remains now engage our attention ranged while living to the shores of an icy sea, and that by some sudden deluge, or vast wave or succession of waves, they were swept from their pasture-grounds. It is not necessary that we should here discuss the extent of this deluge, or inquire whether it covered simultaneously the north of Europe, Asia, and America; or operated by a succession of great waves or more local inundations. What more immediately concerns our subject is, to know that in the drift containing marine shells of existing species, and boulders borne far from their parent cliffs, we have evidence of diluvial action extending from the *ultima Thule* of the American polar sea to far southwards in the valley of the Mississippi.

The identification of the fossil mammoth and rhinoceros of England and Europe with those of Siberia by the first of living comparative anatomists, might lead us to conclude that the same fauna inhabited the northern parts of the new and old world; but I think that we shall find evidence in the bones of bovine animals brought from Eschscholtz Bay, that an American type of ruminants was perceptible even in that early age.

At the present time the moose-deer and mountain sheep inhabit districts of America suited to their habits up to the most northern limits of the continent; while the musk-ox and rein-deer go beyond its shores to distant islands; and the arctic hare is a perennial resident of the most northern of these islands that have been visited, or up to the seventy-sixth parallel. Supposing the climate of North America, at a time just antecedent to the drift period, to have been similar or nearly so to that which now exists, the habits and ranges of the ferine animals at the two dates, though the species differ, may have had a close analogy. The mammoth and other beasts that browsed on the twigs of willows or larger trees may have ranged as far north, at least in the summer, as the moose-deer does now, or up to the seventieth parallel; and lichenivorous or herbivorous ruminants may have extended their spring migrations still further north;—these journeys in quest of seclusion and more agreeable food being quite compatible with the co-existence of vast wandering herds of the same species in more southern lands, reaching even beyond the limits over which the drift has been traced, and where the final extinction of the entire races may be owing to causes operating in comparatively recent periods.

The St. Petersburg Transactions, and other works mentioned in the foot-note*, contain accounts of the circumstances attending the discovery of the entire carcases of a rhinoceros and of two mammoths in arctic Siberia; and one cannot avoid regretting that they were beyond the reach of competent naturalists, who might, by examining the contents of the stomach, the feet, external coverings, and other important parts, have revealed to us much of the habits of these ancient animals and of the nature of the country in which they lived. The inexhaustible deposits of organic remains in the Kotelnoi or New Siberian Archipelago lying off the Sviatoi Noss, may yet disclose some equally perfect carcases; and their exploration by a scientific expedition is a project that promises a rich return for the labour and expense of such an undertaking.

In arctic America such remains have been discovered in its north-eastern corner alone, and as yet, bones, horns, and hair only have been obtained, without any fresh muscular fibre; but all the collectors describe the soil from which they were dug as exhaling a strong and disagreeable odour of decomposing animal matter, resembling that of a well-filled cemetery. In August, 1816, Kotzebue, Chamisso, and Eschscholtz discovered, in the bay which now bears the name of the last-mentioned naturalist, some remarkable cliffs, situated a short way southwards of the Arctic Circle, and abounding in the bones of mammoths, horses, oxen, and deer. The cliffs were described by their discoverers as pure icebergs one hundred feet high, and covered with soil on which the ordinary arctic vegetation flourished. These novel circumstances excited strongly the attention of the scientific world; and when Captain Beechey and his accomplished surgeon Collie, ten years later, visited the same place, their best efforts were made to ascertain the true nature of the phenomenon. Dr. Buckland† drew up an account of the fossil remains then procured, with illustrative plates, and Captain Beechey published a plan of the locality ‡.

This plan comprises a nearly square section of country, having a width and length of about fourteen miles. The Buckland River, where it bends to the northward to fall into Eschscholtz Bay, flanks the district on its inland or eastern border. From the mouth of this river the coast-line trends nearly due west to Eschscholtz Bluff, and forms the south side of that bay; the shore for one-half of the way, or about seven miles, between the Bluff and Elephant Point, being composed of the high icy cliffs, and for the remainder of the distance, or from Elephant Point to the river, the coast is low and slightly incurved. The west face of the land fronts Kotzebue Sound and is formed of slaty gneiss rocks, which terminate on the north at Eschscholtz Bluff, and ten or twelve miles to the southward the rocky eminences, taking an inland direction, are flanked by low, marshy ground. A ridge of hills runs nearly parallel to the western shore at the distance of a mile and a quarter, and at their southern angle, where they bend inland, there stands still nearer the coast-line one of the loftiest bluffs, ascertained to be 640 feet high. From this corner the course of the range is south-easterly, the swampy country above-mentioned running along its base. The banks of the Buckland are also represented as being high, if not hilly,

---

* 1. De Reliquiis Animalium exoticorum per Asiam borealem repertis complectentum, auct. P. S. Pallas. (Novi Commentarii Acad. Scient. Imper. Petropolitani, tom. xvii. pro anno 1772. Petersb. 1773, p. 576.)

2. Notice sur les Bœufs fossiles de Sibérie; par G. Fischer. Bulletin de la Société des Naturalistes de Moscou, seconde année, 1830, No. 1, p. 80.

3. Ueber fossile Reste von Ochsen, deren Arten und das Vorkommen derselben; von Herrmann von Meyer. (Nov. Act. Physico-Med. Acad. Naturæ Curiosorum Cæs. Leop. Carolin. Vratislaviæ et Bonnæ, 1835.)

4. Owen, British Fossil Mammalia.

† Beechey's Voyage to the Pacific and Beering's Straits. Append.

‡ Natural History of Beechey's Voyage to the Pacific.

and they enclose, in conjunction with the range, a sloping valley or basin, drained by numerous rivulets, and opening to the north on the low coast eastward of Elephant Point. At the western entrance of the Buckland there is a minor display of frozen mud-cliffs; similar deposits exist also on its eastern bank as well as on the north shore of Eschscholtz Bay, likewise on various points of the coast between Beering's Strait and Point Barrow; but fossils have been detected only in Eschscholtz Bay, and on the banks of a few rivers that join Beering's Sea between it and Mount St. Elias.

The following extracts from the Narrative of Captain Beechey's Voyage contain a description of the cliffs by a skilful observer.

" We sailed up the [Eschscholtz] bay, [28th July, 1826,] which was extremely shallow, and landed at a deserted village on a low sandy point where Kotzebue bivouacked when he visited the place, and to which I gave the name of Elephant Point from the bones of that animal being found near it. The cliffs are from twenty to eighty feet in height, and rise inland to a rounded range of hills between four and five hundred feet high above the sea. In some places they present a perpendicular front to the northward, in others a slightly inclined surface, and are occasionally intersected by valleys and watercourses, generally overgrown with low bushes. Opposite each of these valleys there is a projecting flat piece of ground, consisting of the materials which have been washed down the ravine, where only good landing for boats is afforded. The soil of the cliffs is a bluish-coloured mud, for the most part covered with moss and long grass, full of deep furrows, generally filled with water or frozen snow. Mud in a frozen state forms the surface of the cliff in some parts; in others the rock appears with the mud above it, or sometimes with a bank half-way up it, as if the superstratum had slid down and accumulated against the cliff. By large rents near the edges of the mud-cliffs they appear to be breaking away, and contributing daily to diminish the depth of water in the bay." (p. 257.)

" Such is the general conformation of this line of coast. That particular formation, which, when it was first discovered by Captain Kotzebue, excited so much curiosity, and bore so near a resemblance to an iceberg as to deceive himself and his officers, remains to be described. As we rowed along the shore the shining surface of small portions of the cliffs attracted our attention, and directed us where to search for this curious phenomenon, which we should otherwise have had difficulty in finding, notwithstanding its locality had been particularly described; for so large a portion of the ice-cliff has thawed since it was visited by Captain Kotzebue and his naturalists, that only a few insignificant patches of the frozen surface now remain. The largest of these, situated about a mile to the westward of Elephant Point, was particularly examined by Mr. Collie, who on cutting through the ice in a horizontal direction found that it formed only a casing of the cliff, which was composed of mud and gravel in a frozen state. On removing the earth above, it was also evident by a decided line of separation between the ice and the cliff that the Russians had been deceived by appearances. By cutting into the surface of the cliff, three feet from the edge, frozen earth similar to that which formed the face of the cliff, was found at eleven inches' depth, and four yards further back the same substance occurred at twenty inches' depth*.

---

* Had the pits been sunk at a distance from the edge of the cliff to the depth of three or four yards, information of a more decided character would have been obtained; for the experiments do not of themselves prove satisfactorily that the frozen mud which was reached so early in the summer as the end of July, at the depth of twenty-two inches, was not merely an unthawed layer of the superficial soil, reposing on pure ice at some distance below.

" This glacial facing we afterwards noticed in several parts of the Sound, and it appears to be occasioned either by the snow being banked up against the cliff or collected in its hollows in the winter and converted into ice in the summer by partial thawings and freezings, or by the constant flow of water during the summer over the edges of the cliffs, on which the sun's rays operate less forcibly than on other parts in consequence of their aspect. The streams thus became converted into ice, either in trickling down the still frozen surface of the cliffs, or after they reach the earth at their base, in which case the ice rises like a stalagmite and in time reaches the surface. But before this is accomplished, the upper soil, loosened by the thaw, is itself projected over the cliff, and falls in a heap below, whence it is ultimately carried away by the tide.

" [September, 1826, p. 323.] The cliffs in which the fossils [collected by Mr. Collie] appear to have been imbedded, are part of the range in which the ice formation was seen in July. During our absence of five weeks, we found that the edge of the cliff in one place had broken away four feet, and in another two feet and a half, and a further portion of it was on the eve of being precipitated on the beach. In some places where the icy shields had adhered, nothing now remained but frozen earth from the front of the cliff. By cutting those parts of the ice which were still attached, the mud in a frozen state presented itself as before, and confirmed our previous opinion of the nature of the cliff."

The above description of these remarkable cliffs has been quoted at length, as it is not only perfectly clear but also concise. The opinions of Captain Beechey and his officers respecting the origin of the ice-cliffs are discussed at considerable length in Dr. Buckland's paper, printed as an appendix to the Narrative of the Voyage. Mr. Collie describes the fossiliferous cliff as facing the north and extending two miles and a half in a right line with few interruptions, and as having a general height of about ninety feet. It is composed of clay, he says, and very fine quartzy and micaceous sand, assuming a greyish colour when dry. The land rises gradually behind the cliff to an additional height of one hundred feet, and is clothed with a black boggy soil, that nourishes brown and grey lichens, mosses, several *Ericeæ, Gramineæ*, and various herbaceous plants, and is intersected by valleys pervaded by streams, and having their more protected declivities adorned with shrubs of willow and dwarf birches. " The specimens taken out of the debris at the foot of the cliff (none were dug out of the cliff itself) were in a better state of preservation than those which had been alternately covered and left exposed by the flux and reflux of the tide, or imbedded in the mud and clay of the shoal. A very strong odour, like that of heated bones, was exhaled wherever the fossils abounded." (p. 509.)

After an interval of twenty-four years, the recent voyage of the " Herald" to this interesting spot has given a third opportunity of collecting fossil bones and examining the structure of these now far-famed cliffs. Captain Kellett, Berthold Seemann, Esq., and Dr. Goodridge, with the works of Kotzebue and Beechey in their hands, and an earnest desire to ascertain which of the conflicting opinions enunciated by these officers was most consistent with the facts, came to the conclusion, after a rigid investigation of the cliffs, that Kotzebue was correct in considering them to be icebergs. I have been favoured with papers on the subject from each of the Herald's officers named above, and shall quote as fully from them as my limits allow, after premising a few general observations on the frozen cliffs of other parts of the arctic coast that have come under my personal observation.

At Cape Maitland in Liverpool Bay, which forms the estuary of the Beghula River, and lies near the seventieth parallel, there are precipitous cliffs from eighty to one hundred feet high, composed of layers of black clay or loam enclosing many small water-worn pebbles and a few large boulders. With the exception of about eighteen inches of soil on the summit, which thaw as the summer advances, these cliffs present to the sea a constantly frozen wall, that crumbles annually under the action of the rays of a summer sun, but the fragments being carried away by the waves and prevented from accumulating, the perpendicular form of the cliff is preserved. Elsewhere on the coast cliffs equally vertical, but having a different exposure, were seen masked by a *talus* of snow, over which a coating of soil had been thrown by land-floods of melting snow pouring down from the inland slopes. The duration of these glacier-like snow-banks varies with circumstances. When the cliffs rise out of deep water, the ice on which the *talus* rests is broken up almost every summer, and the superincumbent mass, previously consolidated by the percolation and freezing of water, floats away in form of an iceberg. In other situations the snow-cliffs remain for a series of years, with occasional augmentations marked by corresponding dirt-bands, and disappear only towards the close of a cycle of warm summers. In valleys having a northern exposure and sheltered by high hills from the sun's rays, the age of the snow may be very considerable; but it is proper to say that though aged glaciers of this description do exist on the shores of Spitzbergen and Greenland, they are of very rare occurrence indeed on the continental coast of America. The ice-cliffs of Eschscholtz Bay may have had an origin similar to that of the Greenland icebergs, and have been coated with soil by a single or by successive operations. I find it difficult, however, to account for the introduction of the fossil remains in such quantity, and can offer to the reader no conjecture on that point that is satisfactory even to myself. The excellent state of preservation of many of the bones, the recent decay of animal matter shown by the existing odour, quantities of hair found in contact with a mammoth's skull, the occurrence of the outer sheaths of bison horns, and the finding of vertebræ of bovine animals lying in their proper order of sequence, render it probable that entire carcases were there deposited and that congelation followed close upon their entombment. A gradual improvement of climate in modern times would appear to be necessary to account for the decay of the cliffs now in progress and the exposure of the bones. The shallowness of the water in Eschscholtz Bay, its narrowness, and its shelter from seaward pressure by Choris Peninsula and Chamisso Island, preclude the notion of icebergs coming with their cargoes from a distance having been forced up on the beach at that place. Neither is it more likely that the bones and diluvial matters were deposited in the estuary of Buckland's River and subsequently elevated by one of the earth-waves by which geologists solve many of their difficulties, for ice could not subsist long as a flooring to warmer water. In short, further observations are still needed to form the foundations of a plausible theory.

Dr. Goodridge describes the several cliffs in succession with much detail, beginning with that next Elephant Point and proceeding to the westward. His paper, though interesting throughout, is too long for transcription entire, and I shall therefore merely abstract the most material parts. He commences by stating that the unusually mild season had produced great landslips and exposed the structure of the several eminences forming the cliffs more extensively than in the year in which Captain Beechey visited them. Elephant Point, forming a high promontory in 1826, had now subsided to a mere hillock by the thawing of the icy substratum, as Kotzebue predicted would happen. A pit was dug to some depth in the loose loamy soil of this hillock, formed of the debris of the ruined cliff, at a point where the thigh-bone of a mammoth protruded above the surface, without any ice being found; but on the east of the hill next in succession, a wedge-shaped landslip had left a triangular chasm, whose floor, elevated twenty feet above the beach, was bounded by walls fifty feet high, of pure transparent ice, and its interior angle, reaching thirty feet backwards from the face of the cliff, exhibited an alluvium seemingly undisturbed since it was originally deposited, and consisting of regular layers of "drift" and peat covered with thick beds of broken sticks and vegetable matter, over which lay a stratum of red river-gravel, then a bed of argillaceous earth, capped by dry friable mould and surface peat, nourishing its peculiar vegetation of coarse grass, moss, lichens, etc. The icy side-walls showed bands or layers considerably inclined, and testifying to their origin in drift snow; and the size of the sticks imbedded in the back walls of the chasm was greater than that of the stems of any of the bushes now growing in the neighbouring ravines. It is to be recollected, however, that a short way up Buckland River, groves of spruce-fir are to be met with. A rivulet separates this hill from Elephant Point, and Dr. Goodridge found some of its slopes to be formed of semi-fluid mud, over which a man could not pass. On the *second* hill or cliff the depth of soil varied with the un-evenness of the ice on which it rested, from twenty feet to less than four, the soil being everywhere dry. On digging in one spot to the latter depth the surface of the ice was found to incline upwards in the direction of the hill, and the soil thrown out by the spade was so pulverulent that it was readily blown away by the wind. The *third* hill, which projected more boldly than the others, contained, as far as it was explored, neither fossils nor ice, but seemed to be entirely composed of thick beds of peat, *logs* of wood, sticks, and vegetable matter, lying generally, but not regularly, in a horizontal position, resting on dry clay, and a bed of river gravel two feet thick. The *fourth* hill presented a higher and more extensive ice-cliff than any of the others, the ice having melted further back towards the centre of the hill, and forming an even wall upwards of eighty feet in height. The *fifth* cliff or marked projection, in proceeding to the eastward, appeared to have sunk bodily from the hill, forming its background, but had left behind it a few icy pillars and detached walls standing twenty feet above the surrounding level surface, and still covered with from seven to ten feet of soil. Water was flowing copiously from these walls of ice, and they were transparent, without admixture of earth, while the soil which capped them was dry and friable. In the slope of this ruined cliff most of the fossils obtained on this occasion were found, a few small fragments only having been gathered from the soft mud at its foot. Some were collected from the surface of the slope, others were dug out at places where the tips of the tusks protruded through the soil.

A deep valley through which a stream of water flows divides the *sixth* hill from the preceding one. Portions of this hill had subsided from the melting of the icy foundation, but in one part a solitary block of ice about twenty feet square rose above the surface, retaining a thin layer of soil on its summit. From the vicinity of this block the hill rose abruptly on all sides; its declivity descended without break to the beach, and its soil, except in the section that had sunk, did not appear to have been ever disturbed. The beach at this place was not composed of muddy detritus, like that which skirted the bases of the other cliffs. A mammoth tusk, having been noticed protruding above the surface of this hill, was traced downwards by digging to the depth of eight feet, and the skull with a quantity of hair and wool were found lying on a thin bed of gravel, beneath which was solid transpa-rent ice. Enveloping the bones there was a bed of stiff clay several feet in thickness, and mixed with them a small quantity of sticks and vegetable matter. The superficial soil was loose and dry. A strong, pungent, unpleasant odour, like that of a newly opened grave in one of the crowded burial-places of London, was felt on digging out the bones, and the same kind of smell, in a less degree, was perceptible in various other places where the cliffs had fallen. From the same pit out of which the mammoth's skull was dug the bones of some smaller animals (a scapula, tibia, etc.) were taken and were duly labelled at the time, but in the course of their transfer from one public department to another, after reaching London, the labels have been lost, together with the specimens of the buried wood, gravel, and other matters found associated with the bones. Dr. Goodridge says that this emi-nence was the last examined, the approach of night having prevented the party from exploring another hill lying between it and Eschscholtz Bluff.   That hill, however, was covered with luxuriant vegeta-tion, and no icy cliffs showed themselves.

" On Choris Peninsula," says the same gentleman, " frozen soil was found at the depth of four feet at the end of September, after an unusually warm summer, and a cask full of flour deposited by Captain Beechey in 1826, on Chamisso Island, was perfectly sound and fit for food when disinterred in 1848.   It was disengaged with much difficulty from the frozen subsoil, and even the iron hoops of the cask were not rusted."   Dr. Goodridge appends to his paper some remarks on the annual waste of the ice-cliffs, and says that the bay is gradually filling up with the clay and soil which are precipitated into the sea on the melting of the ice on which they had reposed*.

* Mr. Seemann describes the ice-cliffs between Elephant Point and Eschscholtz Bluff as being from forty to ninety feet high, and consisting of three distinct layers, the lower one being ice, the middle one clay con-taining fossil bones, and the upper one peat.   "Their sea-face," he says, " when cut perpendicularly, presents a clear view of their structure.   The icy basis is from twenty to fifty feet thick, and decreases every year in extent.   The summer thaw causes the downfall of the two upper layers, and mixes the peat, clay, plants, and ice in a ruinous heap.   The central layer, which varies in thickness from two to twenty feet, contains, inter-mingled with the clay, gravel, sand, and fossil bones of deer, horses, musk-oxen, and mammoths, the whole emitting the heavy odour of a burial-ground.   In 1848," he goes on to say, " we collected eleven tusks, the largest of which, though not entire, weighed 243 pounds, and was eleven-and-a-half feet long, and twenty-one inches round at the base.   The heads of porpoises and antlers of rein-deer were found on the beach, having been deposited there by the natives.   The uppermost layer of peat, being from two to five feet, is destitute of fossils of any description, and bears a vegetation peculiar to moorlands."   Mr. Seemann concludes by expressing his belief that the ice was located in its present site previous to the deposition of the clay and growth of the peaty soil, and is of antediluvian date.

Captain Kellett, in answer to some queries I addressed to him, informed me that the ice-cliffs were in many places as much as sixty feet high, and of pure ice.   He did not think that the ice extended inland as far as the range of hills, though on digging at the distance of a quarter of a mile from the edge of the cliff he found pure ice under a covering of between three and four feet of soil.   In no instance were the fossils imbedded in the ice, but they generally lay on its surface, the large tusks showing through the soil.   Many were gathered from the mud at the base of the cliffs, where they were exposed to the wash of the tide.   In digging within the Arctic Circle to erect marks he always found the soil frozen at the depth of two feet.— Such are the chief particulars that I collected from the three officers quoted above.   The naturalist who wishes to study the subject more deeply will find several opinions discussed in Dr. Buckland's Appendix to Captain Beechey's Voyage, as already mentioned.

NOTICE.

In the following pages the names of the several parts of a vertebra recently introduced into osteology by Professor Owen are generally employed, and very often the terms relating to *position* and *aspect* invented by the late Professor Barclay. The following is the Professor's short explanation of these terms :—

Aspects of the Head, Neck, and Trunk.—An imaginary plane, dividing the head, neck, and trunk into similar halves, towards right and left, is the *mesial plane*. Every aspect looking towards this plane is *mesial*, and every aspect towards the right or left is *lateral ;* every lateral aspect being *dextral* or *sinistral*.

Aspects of the Head.—An aspect towards the plane of the ridge of the occiput is *inial ;* towards the plane of the corona, *coronal ;* towards the base, *basilar ;* or towards the side opposite the inion or ridge of the occiput, *antinial*.

Aspects of the Neck and Trunk.—An aspect towards the region of the atlas, *atlantal ;* towards the sacrum, *sacral ;* towards the back, *dorsal ;* and towards the sternum, *sternal*.

Aspects of the Four Extremities.—An aspect of a bone towards the trunk in the course of the extremity is *proximal ;* and if from the trunk in the course of the extremity, *distal*. (These terms are also employed occasionally in the following pages in regard to the relation of the segments of the vertebral column to the skull.)

Aspects of the Atlantal Extremities.—An aspect towards the side on which the radius is situated is *radial ;* if towards the side of the ulna, *ulnar ;* if towards the olecranon or ancon, *anconal ;* and if towards the side on which the palm is situated, *thenal*.

Aspects of the Sacral Extremities.—An aspect towards the side on which the tibia is situated is *tibial ;* if towards the side on which the fibula is situated, *fibular ;* if towards the side on which the knee-pan is placed, *rotular ;* and towards the side of the ham, *popliteal*.

Common Terms.—*Dermal*, towards the skin ; *peripheral*, towards the circumference ; and *central*, looking towards the centre.

The terms may be employed adverbially, as *coronad*, etc., by substituting *d* for the final *l* or *r*.

(It has been found convenient to employ occasionally in composition the prepositions *ante* and *post* in describing the parts of the spinal column as a series commencing with the first cephalic vertebra.)

LIST OF FOSSIL BONES FROM THE ICE-CLIFFS OF ESCHSCHOLTZ BAY.

THREE several collections made by British officers have been more or less available in furnishing subjects for the following list:—

1st. That made by Captain Beechey and Mr. Collie, which has been described by Dr. Buckland and many of the bones figured on a reduced scale in the 'Appendix of the Voyage to the Pacific and Beering's Strait, performed in the Blossom by Captain F. W. Beechey, R.N., in 1825–28. London, 1831.'

2nd. A collection made by Captain Kellett and Dr. Goodridge, and sent to England under the charge of Lieutenant (now Commander) Wood.

These fossils were deposited in the British Museum, and when I was requested to assist in describing the objects of natural history collected during the Voyage of the Herald, for the illustration of which Government made a liberal grant of money, I applied to the Trustees of the Museum for the loan of them, that I might study them at leisure in connection with the rest of Captain Kellett's collection, and have drawings made of such as it should be desirable to figure. This request was not acceded to, but the trustees gave me permission to inspect them in the Museum, and to send an artist thither to draw them. Mr. Waterhouse, keeper of the fossils, and Mr. Gray, in whose care the recent osteological collection is placed, and by whose exertions it was indeed entirely formed, gave every facility I could desire for availing myself of the permission of the trustees; but my daily official duties at Haslar Hospital prevented me from profiting by the kindness of these officers, except on three or four occasions, when, having obtained leave of absence, I made short and hurried visits to London. Independent of the trouble and expense of transporting the fossils in my own possession backwards and forwards on each occasion for comparison, I did not find these rare visits favourable for the deliberate study such comparisons required, and my arrangements of the bones and determinations of the species are therefore principally founded on a careful study of the

3rd Collection, made by Captain Kellett subsequent to the second one, and brought home by him in the Herald. This collection was sent to the Museum at Haslar Hospital, where it is now deposited, and to it most of the numbers attached to the following list belong.

ELEPHAS PRIMIGENIUS. *Mammoth.*

("Blumenbach," Owen, Brit. Fossil Mammals, p. 218. *Fossil Elephant*, Buckland, App. to Beechey's Voy.)

Teeth.

A medium-sized tusk is now in Haslar Museum, to which it was presented by Mr. Collie, who found it in 1820. Mr. Collie's journal, as quoted in Dr. Buckland's paper, enumerates "nine tusks, five of them large, and weighing from 100 to 160 lbs. each; and four small, one of them found in the debris half-way up the cliff. The largest tusk measured 20 inches in circumference at its root, and 3 feet above the root 21½ inches; another tusk, whose tip is wanting, measures 62 inches along the chord of its curve, and 110 inches following the curve." There were also four fragments of tusks in that collection.

Mr. Seemann informed me that in 1848 the Herald's officers collected eight tusks, the largest of which had lost its point, but still measured 138 inches along its curve and 21 inches in circumference at its base. It weighed 243 lbs. The tusks formed a considerable part of Captain Kellett's first collection, and were, on being brought home by Commander Wood, sent to the Museum of Economic Geology, and afterwards distributed to the British Museum and elsewhere.

Of Captain Kellett's second collection (or the third in order as alluded to in the above remarks) four fragments of tusks were sent to Haslar Hospital, and some more entire ones presented to other institutions. These fragments are in an advanced stage of decay; their laminæ are separated by crystallization of some salts and the formation of a quantity of phosphate of iron.

If we add to the tusks here mentioned those procured by Kotzebue, but whose number I do not know, we must conclude that they were the spoils of at least fifteen mammoths, and collected on four hasty visits, the results of the exposure of comparatively small portions of thawing cliff. The neighbouring Eskimos are in the habit of employing the soundest tusks for the formation of various utensils (see App. Beechey's Voy., plate i. fig. 4); and the American fossil ivory has for at least a century, and for a longer period of unknown duration, been an article of traffic with the Tchutche of the opposite shores of Beering's Strait; so that we can venture upon no calculation of the multitudes of mammoths which have found graves in the several icy cemeteries of the American coasts of Beering's Sea. We may add that from 70 to 100 lbs. are reckoned large weights for the tusks of the living Asiatic elephant, and 325 lbs. and 350 lbs. have been recorded as very extraordinary weights for those of the African species.

No. 12*. A right upper molar, being a perfect tooth, and the largest grinder in Captain Kellett's second collection. It much resembles one from the Essex till (fig. 91, Owen, Foss. Mamm. p. 237), but is not so much decayed. Its greatest length from the anterior convex end to the posterior digital process is 9 inches; the length of its grinding surface 8 inches, and its breadth 3·9 inches. The grinding surface is flattish, with a slight transverse concavity and the ends somewhat worn down, producing a small convexity longitudinally. It consists of twenty-two narrow double plates of enamel, with rather broad cæmentum.

No. 14. A right upper molar, being a rather small but entire tooth. Its crown is much worn down by attrition of the food, so that the plates of enamel no longer form continuous partitions be-

---

* The numbers at the commencement of the paragraphs correspond with the numbers attached to the specimens in the Collection.

tween the dentine and cæmentum; and there are some isolated rings of enamel or sections of digital processes on the mesial side of the tooth towards its middle. The last or 18th plate had been just coming into use. The plates are thin and narrow; four fangs show on the mesial side, and three are well developed on the exterior or lateral side. In length the grinding surface measures 6·5 inches; its greatest breadth is 3·6 inches.

No. 15.   A right upper molar, consisting now of seventeen plates, of which the posterior ones are entire, and the anterior ones broken. This tooth seems to have been chopped and worn by friction between ice and stones, probably after having fallen from the cliff. It has also been worn by attrition in the performance of its functions anteriorly down to the cæmentum, and its last digital process has just reached the edge of the grinding surface. There are six pairs of these processes. A tooth of an African elephant of the same dimensions is composed of ten or eleven plates. Length of the grinding surface 6 inches; its greatest breadth 2·9 inches.

No. 7.   A right upper molar, with nearly all its plates remaining, but the cæmentum in a great part washed out. No fangs remain. Plates seventeen or eighteen in number. Length of grinding surface 6·2 inches; breadth of ditto, having been diminished by loss of cæmentum, only 2·7 inches.

No. 10.   A right upper molar, being only a fragment, consisting of merely the posterior part of the tooth with some of the digital processes broken away. It still contains eleven plates of the broad variety. The cæmentum is much worn and decayed. Length of the grinding surface still remaining 5·5 inches; its breadth 4 inches. These five molars belonged to at least three different individuals, and they were all gathered on Captain Kellett's second visit to the cliffs.

No. 5.   A left upper molar, being the middle part of the tooth only. It is much worn and wasted, especially the cæmentum. The enamel is broken anteriorly, and much of the dentine worn out laterally. There are three pairs of posterior digital processes, the middle ones having the longest antero-posterior diameter and the shortest transverse one. There remain seventeen plates. The length of the grinding surface is 6·5 inches, and its breadth 2·6 inches.

No. 2.   A left anterior upper molar, consisting of seven plates, with broad lamellæ. Grinding surface 3 inches long; 2·2 inches wide.

No. 9.   A fragment of a left upper molar, composed of thirteen plates. Length of grinding surface 4 inches; its breadth 2·8 inches.

Nos. 3, 6, and 1.   Fragments of one crown and two heels of molars.

A left upper molar mutilated, procured on Beechey's voyage by Mr. Collie, is contained in Haslar Museum. Captain Beechey's collection appears to have been rich in molars, of which seven were examined by Dr. Buckland.

No. 11.   Left anterior mandibular molar, composed of fourteen or fifteen plates. Cæmentum much wasted on the mesial side, and between the folds of the enamel. Length of grinding surface 5 inches; its greatest breadth, which is less than when it was entire, is 2·7 inches. A nearly complete mandibular molar of an African elephant of the same shape and size is formed of eleven or twelve plates.

No. 8.   A left mandibular molar, incomplete, but still containing sixteen plates, almost as broad as those of an African elephant. The enamel is broken from decay, and the cæmentum much worn by attrition. Length of grinding surface 7·5 inches; its greatest breadth 3·8 inches.

No. 13.   A right mandibular molar, worn down to the dentine in front, so that the plates are obliterated there; all the posterior ones had come into use. The cæmentum is much wasted, and the grinding surface is more concave and has a more oblique curve than is usually seen in the corresponding tooth of an African elephant. Six inches of grinding surface contain ten plates, before which there is 1½ inches of dentine. The whole length of the grinding surface is 7·8 inches, and its breadth 2·8 inches.

Plate i. fig. 1 and 2, of Dr. Buckland's Appendix to Beechey's Voyage, represent two molars infixed in the mandible. They contain about thirteen plates each, and their grinding surfaces measure by the scale about 6½ inches.

Skull.

Nos. 27, 28, 29, 30, 31, 32, 33, 34, 35, 52. A skull was found with the tusks in their sockets and tips protruding through the soil, as mentioned in page 7 of the preceding geological notices. Unfortunately in digging it out from a depth of eight feet it was much broken, and seventeen or eighteen fragments, being only a small part of the whole, have reached Haslar Hospital. The most recognizable of these are the exoccipitals, which are separated from each other, and somewhat mutilated. The condyles are entire, and measure, from the lateral aspect of one to that of the other when placed *in situ*, more than 8 inches, the greatest width of the occipital foramen being about 3 inches. Detached pieces of the sockets of the tusks have also been preserved, together with parts of the malar and one petrosal. Also the symphysis of the mandible. The tusks were dug out along with those fragments, but have not come under my observation. Neither have I seen a sample of the wool and hair which was found in contact with the skull.

A mandible containing one molar on each side, and wanting only one of the ascending rami, is figured, as has been mentioned above, in Dr. Buckland's paper.

Vertebræ.

No. $\frac{36}{1}$. An atlas, in good preservation, and considerably too small for the reception of the condyles of the skull above noticed, forms part of Captain Kellett's second collection sent to Haslar Museum. The antarticular cavity measures 7 inches across; its sterno-dorsal diameter is 3·5 inches; and the extreme transverse diameter of the vertebra is a little more than a foot. This bone retains much of its original weight, is so little absorbent that the tongue adheres very slightly to it, and the articular cavities are still glazed with their smoothened surfaces. Some parts of the neural arch are decayed, and phosphate of iron has formed in one place.

No. $\frac{36}{2}$ is a dentata or second cervical, which has lost most of its animal substance, and is much worn down. This was probably picked up on the beach, having evidently been rolled among the gravel and much wasted thereby. It is very light.

No. 172. Fragment of a seventh cervical, composed of the left prozygapophysis and zygapophysis. It is pretty compact, and is partially stained by phosphate of iron.

No. 41 is a nearly entire dorsal. The lateral processes have suffered some loss, and the antarticular epiphysis of the centrum has dropped off: but the neural spine is perfect, and the zygapophyses retain their smooth articular surfaces, as in a recent skeleton. Outline of the post-articular cup of the centrum ovate or heart-shaped, the neural canal forming an obtuse notch in the broad end.

Transverse diameter of the centrum on the sternal side of the distal cups for the head of the ribs	5·9 inches.
Transverse diameter, dorsad of the same cups	6·6
Distance between the sternal surface of the centrum and the neural canal	4·8
From sternal surface of centrum to crown of neural arch, distal end	6·6

Atlanto-sacral axis of centrum, excluding detached epiphysis   .   .   .   .   .   2·8 inches.
Extreme width at the lateral processes   .   .   .   .   .   .   8·8
Length of neural spine from the atlantal edge of the prozygapophysis .   .   . 10·0
From the sacral edge of the zygapophysis the spine measures only   .   .   .   5·5

The difference is the extent of the neural canal protected dorsad by this vertebra. This bone is in nearly the same state of preservation, and is of the same colour with the atlas $\frac{36}{1}$, and may have belonged to the same individual. It appears to have been the twelfth or thirteenth dorsal of a pretty large but not aged individual, as the articular epiphyses have not firmly coalesced with the centrum. The skull belonged to a larger animal.

No. 40. This is the centrum of apparently the first dorsal of an individual of nearly the same size as the one to which the preceding segment of the vertebral column belonged, but it has been much mutilated by decay, and it has lost more of its animal matter, since the tongue applied to its surface adheres strongly. The transverse diameter of the distal articular surface is about 5 inches; the sterno-neural diameter about as much; and the length of the centrum, exclusive of some loss by waste, is 2·9 inches.

No. 39. A fragment of a second lumbar, that may have belonged to the same individual with the preceding dorsal. The bone is equally decayed.

Breadth of distal articular surface of the centrum   .   .   .   .   .   5·2 inches.
Sterno-neural diameter of ditto .   .   .   .   4·0
Transverse diameter of the neural canal   .   .   .   .   .   .   3·6
Sterno-neural axis of ditto .   .   .   .   2·1

No. 37. Last dorsal, with the centrum and sides of the neural canal nearly perfect, the rest of the processes mutilated or entirely broken away.

Breadth of the articular surfaces of the centrum .   .   .   .   4·0 inches.
Sterno-neural diameter of ditto .   .   .   .   .   .   3·4
Length of centrum at its lateral margins, where it is greatest   .   .   .   2·7
Length of axis of the centrum between the centres of its articular surfaces .   2·5
Width of neural canal   .   .   .   .   2·0
Sterno-dorsal diameter of neural canal   .   .   .   .   .   1·5

No. 38. A first lumbar, agreeing in size, colour, and condition with the preceding one so closely as to leave little doubt of their being portions of the same skeleton. Their small size shows that the animal to which they belonged was young; but they bear evidence of having been the subjects of inflammatory action for a length of time, in their surfaces everywhere, except the articulations, being rugged with exostoses.

Mr. Collie's list, quoted by Dr. Buckland, enumerates three dorsals of the mammoth, each 5½ inches in diameter. If this measurement refers to the articular surfaces, the vertebræ must have been bigger than any found by Captain Kellett's parties.

Pelvis.

No. 61. The upper part of the acetabulum and a portion of the front of the ilium, much decayed.
No. 53. A fragment of the ischium, and of the edge of the acetabulum.
Captain Beechey brought home an imperfect ilium, a fragment of the pubal, and a nearly perfect *os innominatum*, which is represented in Dr. Buckland's second plate, fig. 9, reversed. The breadth of the iliac plate measured on the figure is 44½ inches, and the diameter of the acetabulum 7·4 inches.

Extremities.

No. $\frac{28}{10}$. A considerable portion of the right scapula, being part of the collection made on Captain Beechey's visit to the ice-cliffs, was presented by Mr. Collie to Haslar Museum. This bone is much decayed, and had apparently lain long, together with most of the fossils then obtained, in the mud-bank washed by the tides. Its glenoid cavity has a long axis of 8 inches, a shorter one of 4·5 inches, and the entire length of the fragment is 27 inches.

Another incomplete scapula obtained at the same time, and measuring 31 inches in total length, is probably the left scapula of the same individual. It is figured by Dr. Buckland in his second plate (fig. 8), and is mutilated in about an equal degree as the preceding one. Two other fragments are enumerated in Mr. Collie's list.

No. 55. A more mutilated left scapula exists in Captain Kellett's second collection. Its glenoid cavity has a long axis of 8 inches, and a short one of nearly 4 inches. Much of the animal matter of the bone has perished.

No. 58 is also a part of the same collection, now in Haslar Museum, and is a portion of the blade and root of the spine of the left scapula of a larger animal, being much thicker than that of any of the scapulæ mentioned above. It is much worn, and has suffered friction on the beach, but is still heavy.

No. 60 is a fragment of the blade of a right scapula, still more compact and heavy.

No. 173. The triangular process of the spine of a right scapula, in the same state of preservation with Nos. 58 and 60.

No. 56 is the head of a right scapula of a small mammoth, much mutilated. The shorter diameter of the glenoid cavity is 3·1 inches.

No. $\frac{44}{1}$ is the mutilated head of a left humerus. Its articular surface has a diameter of 5·4 inches.

No. $\frac{44}{2}$. A fragment of the shaft of a left humerus, more compact and heavier than the preceding.

The portion of a humerus brought home by Captain Beechey, and now in the British Museum, belonged to a larger animal than No. $\frac{44}{1}$ did.

No. 42 is the upper part of the shaft of a right radius, part of Captain Kellett's second collection. It is the bone of a small-sized animal, probably about four years old, and though mutilated at the ends, has not lost much animal matter. Its surface is in good condition, and little absorbent. There are two pairs of grooves on the shaft, which have the appearance of having been made by the teeth of some rodent while yet fresh.

No. 57. Part of the shafts of the radius and ulna, coalescent where they cross each other, and much ground down by friction among the gravel.

No. 42. A left radius, which has lost the distal articulating surfaces and about 2½ inches of its shaft.

Length of the fragment	19·5 inches.
Circumference of its middle	6·2
Transverse diameter of proximal articulating surface	2·1

No. 66. Right semilunar bone, nearly perfect, being slightly abraded on the edges only.

The transverse measurement of the palmar or fore side is	4·0 inches.
Its height is	1·9

No. 67.   The left unciform bone, equally perfect with the preceding, and perhaps belonging to the same individual.

Transverse diameter of the palmar or fore side   .   .   .   .   4·2 inches.
Height at the mesial edge, where it is greatest   .   .   .   .   3·2
Antero-posterior axis of the articular surface for the outer metacarpal   .   .   .   3·3
Greatest lateral breadth of same surface   .   .   .   .   1·6
Antero-posterior axis of the triangular articulating surface of the fourth metacarpal   .   3·5
Greatest lateral breadth of the same surface at the base or anterior side of the triangle .   3·0

This articulating surface is separated from the former by a nearly obsolete groove, and from the inclined surface, which receives part of the head of the third metacarpal, by an obtuse ridge.

No. 47.   The shaft of a fœtal? femur.   The articulating cartilages of both ends have separated and the trochanter is gone.

Length of the fragment   .   .   .   . 11·0 inches.
Circumference of the middle of the shaft   .   .   .   . 4·6

Nos. 48, 49, 50, 51, 53, 155, 163, and 193, are fragments of the shafts of various femora, some of the pieces being very compact and weighty, though brittle.   A piece of one of these (No. 48) which is very heavy, though one of the most absorbent in the collection when the tongue is applied to it, being immersed in diluted hydrochloric acid until the earthy matter was removed, swelled out into thrice its original bulk, and appeared to be composed of groups of Haversian cells.

No. 63.   A portion of the head of a femur.   Some of the cancelli contain ochre and phosphate of iron.

No. 64.   Part of the condyle of a femur, much decayed.

Mr. Collie's list enumerates five femora, one of them nearly complete.   The one figured by Dr. Buckland (pl. ii. fig. 5) wants the trochanters.   By the scale appended to the drawing the total length of the bone is 41 inches.   The bone itself is in the British Museum, and has a larger head than No. 63.

No. $\frac{46}{1}$.   The upper part of the shaft and part of the proximal articulation of the left tibia.   This is smaller than a more complete tibia, which forms part of Captain Kellett's first collection, and is now in the British Museum.   The width of the knee-joint of No. $\frac{46}{1}$ is 6·2 inches.

No. $\frac{46}{2}$.   The same parts of the tibia of a much smaller animal.   The width of the knee-joint is under 4 inches.

Nos. 45, 54.   The distal ends of the shafts of two tibiæ, a right and a left one, wanting the ankle-joints, and corresponding in size nearly to No. $\frac{46}{1}$.

No. 156.   A smaller fragment of the shaft of a tibia.

No. 62.   The distal articular surfaces of a larger tibia than the one alluded to above, as forming part of Captain Kellett's first collection, and bearing, in the British Museum, the number 24,581.

No. 68.   An astragalus, much abraded.   A more perfect one exists in the British Museum, brought from the same quarter.

Mr. Collie's list mentions one cuboid, one cuneiform, one tarsal, and one phalangal bone.   Also an entire os calcis taken out of the cliff.   This has been figured by Dr. Buckland (pl. ii. fig. 10).

In referring all these bones to the Elephas primigenius I have not been guided by actual comparisons with authentic bones of that species, not having the means of access to such, but have acted upon the opinion of Professor Owen, who, in his excellent work on British Fossil Mammals (p. 244), refers the tusks brought by Captain Beechey from Eschscholtz Bay, the great mammoth skeleton from the Lena, which is now the ornament of the St. Petersburg Museum, and numerous remains of fossil elephants scattered over England and the British Channel, all to that species.

EQUUS FOSSILIS. *Fossil Horse.*

Mr. Collie's list of the fossils of Eschscholtz Bay, collected on Captain Beechey's visit, mentions only three bones of the horse, viz., an astragalus, a metacarpal, and a metatarsal. The number obtained by Captain Kellett at his second visit was greater, and we have carefully compared them with the corresponding bones of a mature cart-horse of full but not extraordinary size. They are all much larger than those of an aged Shetland pony in the same museum, with which they have also been compared, and may have belonged to a horse about 15 or 15½ hands high.

No. 144. A tolerably perfect sacrum, with the segments so completely coalesced as to leave no trace of the separations of the centra. The fifth centrum has broken away at the joint, but its spine had coalesced with that of the fourth. In the sacrum of a recent horse the fifth has coalesced as completely with the fourth as the preceding ones of the series have with one another. The differences of form between the recent and fossil sacra are very slight. All that I have been able to detect are that the foramina of the third pair are, on the pubic side, further apart in the fossil; that the prozygapophysial articulating surface is somewhat concave laterally, and not so convex in the other direction; that the surface which articulates with the ilium is less extensive and not so continuous; and that the neural spines are more oblique. The dorsal foramina are proportionally larger than in the recent bone, and the deep furrows, which are prolonged from their distal margins, are indistinctly represented in the recent horse.

	SHETLAND PONY.	FOSSIL.	CART-HORSE.
Length of first four sacrals	3·5 in.	6·0 in.	6·0 in.
Width of the first sacral on the pubic side, in a line with the proximal edges of the foramina	2·9	3·5	4·2
Length of zygapophysial process of same sacral, from the lateral border of the foramen on the pubic side	2·4	3·8	4·2
Width of centrum of third sacral at its distal end	1·1	2·2	3·0

The summits of the neural spines of the fossil are broken off. The sacrals diminish more rapidly in their extreme breadth in the horse than in the ass.

From the dimensions recorded in the above table we may gather that the sacrals of the fossil horse are proportionally longer than those of the existing animal. The sacrum and the other bones of the horse which follow have lost a considerable quantity of their original weight, and the tongue adheres to them pretty strongly.

No. 100. Right *os innominatum* with the proximal edges of the expansion of the ilium and the tuberosity of the ischium broken off, otherwise perfect and in good condition, though with considerable loss of animal matter. It fits sacrum 144, and from the similarity of their colour and their state of preservation they may be considered as belonging to the same individual. The sacro-pubic diameter of the pelvis is 6·4 inches; that of the cart-horse, with which it was compared, being 9·0 inches, and that of the Shetland pony 4·8 inches. The acetabulum is larger in the larger animal,

D

but there is no other difference. The muscular grooves in the ascending ramus of the ilium are more strongly impressed in the domestic animal, as are also the two pits above the brim of the acetabulum, and are to be looked upon as having been augmented by labour in harness. In the larger curves of the bones there are no striking discrepancies. The flat surface on the mesial aspect of the ramus of the ilium is proportionally broader in the fossil, the sternal or anterior side of the ramus proportionally narrower, and the lateral one more nearly equal to that of the cart-horse.

The *foramen ovale* is longer and more regularly oval in the fossil, its diameter being increased by the greater narrowness of the pubal. The greater breadth of the pubals in the cart-horse may be attributed in some degree to the state of servitude in which it had passed its life, for its pubals are very concave on their sacral aspect, evidently resulting from the pressure of a distended bladder, while they retain in the fossil the convex, somewhat spirally twisted appearance, which they present in the aged Shetland pony.

The greater diameter of the *foramen ovale* gives also a longer descending ramus of the ischium to the fossil, both relatively and positively.

	SHETLAND PONY.	FOSSIL.	CART-HORSE.
Circumference of the ramus of the ilium	2·9 in.	4·3 in.	5·1 in.
,, ,, ischium	2·4	3·4	4·4
,, ,, pubal	2·2	2·9	3·8
Longest diameter of the foramen ovale	1·9	3·2	3·1
Transverse ditto ditto	1·2	2·0	2·4
Breadth of ischium immediately below foramen ovale . .	1·5	2·9	3·2
Longest diameter of the acetabulum	1·8	2·6	2·9

No. 101. The acetabulum and descending ramus of the ischium of the right side of the pelvis, scarcely appreciably larger than the preceding fossil, and agreeing in all respects with the same parts of that specimen.

Extremities.

No. 112. The radius of the right side with the lower portion of the ulna coalesced, and the olecranon broken away. From the colour and condition of this bone and its relative size it may have belonged to the same skeleton with the *os innominatum* described above.

	CART-HORSE.	FOSSIL.
Extreme length	14·4 in.	13·4 in.
Lateral diameter of the narrowest part of the shaft	1·8	1·6

The resemblance between the fossil and recent bones is very close, the principal difference lying in the articulating surface of the elbow-joints ; the cavity that receives the outer or lateral process of the ulna being more incurved in the fossil, and rendering the cup for the reception of the lateral condyle of the humerus narrower in its antero-posterior axis. In the tarsal articulation of the bone also the ulnar or lateral half of the joint is proportionally wider than in the cart-horse, the mesial or radial portion being narrowed in an equal proportion.

No. 96 is the lower part of the tibia of the right leg, and is in the same state of preservation with the radius and pelvis above noticed.

Lateral diameter at the tarsal joint	2·8 inches.
Ditto of the shaft where narrowest	1·7
Antero-posterior diameter of the shaft at the same place	1·3

No. 95 is also a fragment of a right tibia of the same dimensions with the last, but of a darker colour.

No. 97 is a slightly larger right tibia, nearly perfect, and exactly fitting astragalus 142. It is of a darker hue than the preceding tibia, agreeing in that respect with the astragalus, and raising a presumption that they were obtained from the same spot. It is of equal length with, but wider at the joints than, the tibia of a musk-cow.

	FOSSIL.	CART-HORSE.
Extreme length measured from central process in the knee-joint . . .	14·1 in.	15·7 in.
Transverse diameter of knee-joint	4·0	4·2
Circumference of the shaft at its most slender place in its lower third . . .	5·2	5·4

A careful comparison of the fossil tibia with that of the cart-horse did not enable me to detect any decided difference of form. The tibial spine, in form of a subacute line, descends in the recent tibia nearly to the lateral side of the distal joint; in the fossil it cannot be traced quite through the proximal half of the bone, but this may be in part attributed to some friction and waste of surface. On the popliteal aspect of the bone, near the ham, the muscles have made deeper impressions in the fossil, and the middle of the shaft is less convex on that side in the cart-horse.

No. 94. A fragment of a right tibia, with the knee-joint and fibula broken away. It is of a reddish hue, has a more compact texture, and its surface is more perfectly preserved than the preceding tibia. In it also the lines and grooves formed by the muscles are more numerous and acute than in the cart-horse. The tongue adheres but slightly to this bone in comparison with No. 97. The astragalus No. 142 fits this tibia nearly, but not exactly, and it is of a different colour, being blackish-brown.

No. 93. The lower part of a left tibia, nearly of the same size with the preceding ones, but not pairing exactly with any of them. In the size of its distal joint it comes nearest to No. 94, but its shaft is perceptibly more slender; it is of a paler colour, and its surface is more weathered. The tongue adheres strongly to it.

There is little doubt of the five fragments of tibiæ enumerated above having belonged to five different individual horses; and when we consider that the mammoth teeth are more attractive to collectors, we may infer that the horses were little less numerous than the mammoths at the time when their carcases where swept into the cliff-cemetery.

Nos. 142, 143. Two right astragali mentioned before, and strongly resembling that of the recent animal; the oblong and somewhat kidney-shaped articulating surface which plays upon the interior or mesial process of the heel-bone alone presenting some difference in outline. No. 142 fits tibia 97 exactly, as noted above. No. 143 cannot be made to fit any of the fossil tibiæ, though the difference of size is trifling. This extends the number of individual skeletons that furnish the horse bones in the collection to six.

No. $\frac{142}{2}$. Fragment of the distal end of a right metatarsal, less than the metatarsal of the cart-horse in a degree corresponding with the general inferior size of the fossil bones.

The existing race of American horses is generally considered to have been introduced into the continent by the Spaniards, since the invaders found the Mexicans everywhere unacquainted with the animal. At the present time herds of wild horses are limited to the prairies, and are not numerous beyond the confines of New Mexico, nor do they range northwards beyond the 49th parallel of latitude. Domestic horses have only very recently been carried out of Rupert's Land to the banks of the Mackenzie, and I believe that none have as yet been conveyed to the northern parts of Russian America.

CERVUS ALCES.   *The Moose-deer.*   (Muswa of the Cree Indians.)

No. 114. This is a fragment of the frontal and left antler of a Moose-deer, and No. 113 is a detached piece of the palm of the antler. The antler had acquired a bony hardness on the exterior only, and that partially, while its interior is spongy, but it seems to have attained its full size; we may therefore conclude that the animal's death took place towards the end of summer, or in the month of August. The bone has the dark hue of the other fossils, but is compact and weighty, having lost little of its animal substance. Its cells are filled with the same micaceous sandy loam that formed the general matrix of the bones, and phosphate of iron has formed in minute grains in many parts of the broken surface. Were it not that the moose-deer still ranges to the Arctic Sea, and that its bones are not enumerated among the spoils of animals of the drift period dug up elsewhere, I should have had no doubt of these fragments of the skull of a moose being equally fossil with the other portions of the collection. I have had no opportunity of comparing the large deer bones mentioned by Dr. Buckland with those of the moose, as this would have settled the matter. In No. 114 the process of the frontal which supports the antler has a circumference of 8 inches, and the stem of the antler itself just above its basal tuberculated ring of 7½ inches. The frontal 1½ inch thick. The animal was not old, as the sagittal suture had not closed between the antlers.

No. 198 is a fragment of the shaft of a tibia, more decayed than the frontal. It is too small to determine the species.

The tibia and radius figured by Dr. Buckland in the Appendix to Beechey's Voyage (pl. iii. fig. 12, 13) seem to have been part of the skeleton of a moose. According to the scale on which the figures have been engraved, the ulna measures 16·2 inches in length, and the radius 13·4 inches.

In the present day the moose ranges from the valley of the St. Lawrence to the Arctic Sea, keeping in the woody districts, being little known on the prairies, and never seen in the " barren grounds" of the north. It frequents willow thickets, and follows these along the banks of rivers, beyond the woods, up to the 69th and 70th parallels.

CERVUS TARANDUS.   *Fossil Rein-deer.*

The rein-deer is at the present day an inhabitant of the country round Eschscholtz Bay. In Rupert's Land and the northern extremity of the continent east of the Rocky Mountains three races of rein-deer are known and recognized by the natives and fur-traders, all passing under the French Canadian appellation of Carriboo (*Quarré-bœuf*), derived from the size of their antlers. The smallest is the barren ground rein-deer, which brings forth its young in the islands and on the coasts of the Arctic Sea, and does not migrate further south in winter than to the skirts of the woods; the largest inhabits the wooded valleys of the Rocky Mountains, bordering on the Mackenzie; and the third race, of an intermediate size, frequents the wooded and hilly districts of Rupert's Land, passing during winter into the interior, and migrating in summer to the coasts of James's Bay. This kind seems to have been formerly plentiful as far south as the State of Maine, and small herds still frequent the borders of Lake Superior and many parts of Canada. No comparisons have been made of the osteology of these several races, and I do not know whether it be the small barren ground rein-deer or the Rocky Mountain one that frequents Kotzebue Sound. The first two cervicals, a radius and ulna, and a femur, all recent bones, of one picked up near an Eskimo village, were brought home by Captain Kellett, but whether they are the bones of a female or male is unknown. They have been exposed long to the atmosphere, but are in a very different condition from the fossils enumerated below, and may be readily recognized as recent bones. The extreme length of the radius is 10·7 inches. The olecranon rises 2·3 inches above the crescentic articular surface of the ulna, or 3 inches above the connection of that bone with the radius. The femur is 11·4 inches long, 2·4 inches wide at the knee, and nearly 3 inches in circumference at the middle of the shaft. It is to these bones and various antlers that Mr. Seemann alludes in his remarks. None of them were put up along with the fossils.

Nos. 107, 108, 109, and 110. Fragments of antlers of rein-deer, evidently fossil, being more or less decayed, and as absorbent when the tongue is applied to them as any of the mammoth's bones.

No. 137. Fragments of the pelvis, consisting of the ascending stem of the right ilium, parts of the ischium and pubal, and the entire acetabulum, which exactly fits the head of the recent femur mentioned above. The bone has lost much of its animal substance, and, with the antlers and other specimens which follow, has the dark colour and fossil aspect of the other bones in the collection. The fine-grained micaceous matrix still adheres to them, and fills the cells of their broken ends. The greatest diameter (or vertical one) of the brim of the pelvis is 1·4 inch; the transverse diameter is 1·3 inch.

No. 167. Part of the right scapula, including the articulation. Longest axis of glenoid cavity, 1·45 inches; shortest one, 1·05 inch.

No. $\frac{112}{2}$. Metacarpal bone, in good preservation, dark-coloured, but not very absorbent.

Its extreme length is	8·0 inches.
The transverse diameter of its proximal joint . .	1·3
That of its distal joint	1·6
And of the middle of its shaft	0·8
Depth of its posterior groove where the edge is highest . . .	0·42

No. $\frac{111}{2}$ Part of the shaft of a tibia.

No. $\frac{111}{1}$ Metatarsal bone, in the same condition with the metacarpal. It is very little longer than the fossil metatarsal of a rein-deer found in the fens of Cambridgeshire, and figured by Professor Owen in his Fossil Mammals.

Its extreme length is	10·9 inches.
The transverse diameter of the proximal joints	1·2
Antero-posterior diameter of ditto	1·3
Transverse diameter of the distal joint	1·7
Greatest antero-posterior diameter of the shaft about 2 inches from the tarsal joint .	1·6
Greatest depth of the posterior groove . .	0·6

A metatarsal of a female rein-deer skeleton in the College of Surgeons measures 10 inches in front, excluding the interior projections of the joints.

The antler figured by Dr. Buckland in the Appendix to Beechey's Voyage (pl. iii. fig. 11) is much like one of the fragments brought home by Captain Kellett.

OVIBOS MOSCHATUS. (Ovibos Pallasii, *Dekay.*)   *Fossil Musk-ox.*

Dr. Buckland in his paper on the ice-cliff fossils has the following passage :—" Pl. iii. fig. 1. Head of a *Bos urus,* in precisely the same condition with the fossil bones of elephants, and very different from the state of the head of a musk-ox, with the external case of the horns still attached to it, which was brought home with the fossil bones, and was found with them on the beach at the bottom of the mud-cliff in Eschscholtz Bay, but is so slightly decayed that it seems to have been derived from a carcase that has not long since been stranded by the waves.   This head of a musk-ox is not engraved, as it cannot be considered fossil."   Dr. Buckland has not here considered the preservative powers of constant frost, and his conclusions as to the musk-ox cranium being derived from a recent animal are invalidated by the remains of musk-oxen undoubtedly fossil having been taken out of the stratum in which the fossil bones are deposited.   It is fair, therefore, to conclude that the cranium had fallen with a portion of the cliff.   The horn-cases of the fossil bison are found in the deposit still more fresh than that of the musk-ox.

No. 24,591, 1 B, are the numbers attached to the skull of the musk-bull referred to by Dr. Buckland, and deposited in the British Museum.   It had belonged to an aged animal, its orbital plates being very thick and rough, as they are in the old bulls.   The bones of the face are broken away.

No. 103.   A fragment of the frontal and right horn-case of a musk-bull, as much decayed and having lost as much animal substance as the mammoth bones.   Its cavities are filled with the micaceous loam in which the fossils are imbedded in the undisturbed cliffs.

No. $\frac{126}{2}$ is the atlas of a musk-cow, with the transverse processes broken away.   It has a smaller and less prominent and smoother hypapophysial tubercle than the atlas of the recent musk-bull, and its neural tubercle or abbreviated spine is not only less prominent but is moreover divided by a longitudinal furrow.

Transverse diameter between the outer edges of the prozygapophysial articulating surfaces	4·5 inches.
Transverse diameter of the neural canal	1·5
Sterno-neural diameter, including furrow in the centrum	1·2
Length of centrum on sternal aspect	1·6
Width of notch which receives the shoulders of the basi-occipital	2·3

The length of the centrum and neural canal of the recent musk-bull (pl. v. fig. 1) is considerably greater.

No. 128.   A third dorsal of a musk bull or cow, well preserved (the tip only of the neural spine being broken off), and undoubtedly fossil.   The centrum retains much of its weight, and is very compact, but it is absorbent, and the neural spine, which is more decayed, is still more adherent to the tongue.   The antarticular surface of the centrum has a very moderate convexity, and the postarticular cup is shallow.

Length of centrum excluding articular surfaces	1·7 inches.
Sterno-neural diameter of proximal articular surface of centrum	2·0
Ditto ditto of distal ditto	1·8
Lateral diameter between the outer edges of the proximal cups for the heads of the ribs	2·0
Ditto ditto ditto distal cups for ditto	3·1

Lateral diameter between the outer sides of the diapophyses . . . . . . 3·5 inches.

Ditto   ditto   ditto   of the zygapophyses . . . . . 1·5

The neural canal, proximal end, is almost circular, with a diameter of about 0·8 inch.

No. 135 is the mutilated fifth lumbar of a musk-ox, most of the processes being more or less injured. It is decayed, has lost much weight, and is of a dark colour. In size and character of the remaining parts it corresponds exactly with the same bone in the skeleton of the musk-bull.

No. 102. Sacrum of a musk-bull, consisting of four vertebræ, well preserved, but in a similar condition and colour to the bulk of the other fossils. The neural canal was filled with micaceous loam.

	FOSSIL.	RECENT MUSK-COW.
Transverse diameter of the proximal articular surface of centrum . . .	2·7 in.	2·4 in.
Height of neural spine above the crown of the pointed neural arch, proximal end	1·7	1·6
Greatest width between prozygapophysial articular surfaces	2·6	2·3
Extreme width of proximal lateral processes	6·6	5·9
Length of chord of four sacrals, pubic side	6·3	5·5

No. 136. Part of the acetabulum and ramus of the right ilium of a musk-ox, similar in colour and loss of animal substance to No. 102.

No. 206. The condyles of the humerus of a musk-ox, very much decayed. In Captain Kellett's first collection also there is a part of the humerus of a musk-ox.

No. 98. Proximal end of the left tibia of a musk-ox, of a very dark colour, and in part decayed, but still retaining more animal matter than some of the other bones of this species enumerated above.

	FOSSIL.	RECENT COW.
Extreme width of the knee-joint	3·5 in.	3·4 in.
Antero-posterior axis of joint	2·3	2·2
Distance between the most prominent part of the proximal end of the tibial spine to the posterior notch at the inner margin of the surface that articulates with the left femoral condyle	3·1	3·0

With regard to the present range of the existing musk-ox, our knowledge of it may be considered as pretty correct, except in respect to Russian America. It has not been seen alive on the Labrador coast, in Greenland, nor in the insular lands lying between Hudson's Straits and Lancaster Sound. It ranges on the continent from the limits of the woods northwards, and from Seal River, in the Welcome, to the eastern outlet of the Mackenzie, seldom in winter penetrating into the thick forests, but rather resorting to the thin and scattered clumps of spruce-fir that skirt the barren grounds; and in summer migrating partially northwards to the islands beyond Barrow's Strait, while the bulk of the herds remain all summer on the barren grounds, chiefly in hilly districts not far from the woods. Its southern limit may be nearly traced by a line running from the entrance of the Welcome into Hudson's Bay, about the 60th parallel of latitude, in a westward and northward direction to the 66th parallel at the north-east corner of Great Bear Lake, and from thence in nearly the same direction to Cape Bathurst, in the 71st parallel. Hearne states that he once saw the tracks of one within a few miles of Fort Churchill, in latitude 59°; and this seems to have been its southern limit also as far back as the time when M. Jeremie was at Churchill. The last-named author is the first who mentions the animal, and he assigns the country lying between the Danish or Churchill River and Seal River as its habitation. The accounts to be found in the older writers of its having been seen further south have chiefly arisen from the American bison having been mistaken for it. The natives report a ruminant as big as, or bigger than, the musk-ox as an inhabitant of the barren lands west of the Mackenzie and north of the Arctic circle, and I have recently heard a repetition of this rumour, which I received many years ago in a very vague shape. I have not been able, however, to obtain any authentic description of this animal, nor have I learnt that utensils made of the musk-ox horn are common on the Arctic coasts to the westward of the Mackenzie. Such articles pass from one Eskimo horde to another in traffic, and their occurrence in any quarter is not therefore decisive of the musk-ox being an inhabitant of the neighbourhood; but as the spoils of the animal are often met with in the Eskimo encampments east of the Mackenzie, we should expect to find them also west of the Mackenzie, if the living animal ranged so far. Neither Sir John Franklin, nor Dease and Simpson, nor Commander Pullen, seem to have seen anything of the kind in their voyages along the Arctic coasts of Russian America. Fabricius mentions the skull of a musk-ox as having been found on the Greenland coast, and Pennant thinks that it was carried thither on ice.

More recently a fragment of the skull of a musk-ox has been found in the alluvium of the valley of the Mississippi, at New Madrid, to the southward of the 37th parallel, and below the junction of the Ohio with the Mississippi. It is described by Dr. Dekay under the appellation of *Bos Pallasii.* (Ann. Lyc. New York, p. 11, t. 6.)

Pallas had long before described one fossil musk-bull cranium found under the Arctic circle, on the banks of the Ob, and another found in a "tundra," or barren marsh or moor, still further north (Novi Comm. Acad. Scient. Imp. Petropolit., tom. xvii. tab. xvii. fig. 1–3, pro anno 1772); and in 1809–10 M. Ozeretskovsky described in the Mém. de l'Acad. de Pétersb., t. iii. p. 215, pl. i., a third fossil skull from Siberia, found at the mouth of the Yana, between the Lena and Indigirska. The figures of the latter are copied on a greatly reduced scale by Cuvier in the Oss. Foss. vol. iv. pl. xi. fig. 6, 7, and represent an adult but not an aged male skull. From the horn-cases being represented in fig. 7, one may suspect that some additions have been made by the artist to the correct drawing.

Baron Cuvier thinks that these figures of M. Ozeretskovsky represent a skull with a shorter nose, and differently proportioned ali-sphenoid from those of the living species. On endeavouring to put the recent skull into the same position with that in which the fossil had been placed, I find that in representing the base of the skull the palate must be looked at very obliquely, with the premaxillaries much raised, to bring the orbits in front of the horn-cases, as they are shown in fig. 13 of the Russian author, or fig. 6 of the Oss. Foss., and that consequently the roof of the mouth was much foreshortened, as Cuvier had suspected.

Pallas figures the skull obtained on the Ob, and gives a table of measurements, which I have contrasted with the dimensions of the same parts, as well as I could determine them, of the young but adult male, hereafter described. Pallas does not specify the scale he employs, but in a preceding paper in the same volume the London duodecimal scale is mentioned. For convenience I have turned the lines into decimals of an inch*.

---

* Cuvier, in his Oss. Foss., p. 157, says that Pallas employed the Paris foot as his scale of measurement, but I did not find it so noted in the paper, and have put down Pallas's numbers as I found them, with the exception of turning the lines into decimals of an inch.

	FOSSIL.	RECENT.
Distance between the hinder borders of the bases of the horn-cores and the anterior end of the molar alveoli	— in.	18·2 in.
Distance from ditto to the posterior end of the alveoli	14·3 ?	12·7
„ from the occipital foramen to the anterior border of the molar alveoli .	15·1 ?	13·7
Longitudinal axis of the base of the horn-core	6·7	9·0
Breadth of each base of the horn-core	2·9	5·0
Distance between horn-core and nasal suture	2·5	2·5
Projection of the same behind the occiput	0·5	1·7
Length of the depending horn-core processes from their bases . . .	5·7	6·5
Their breadth at the base	3·7	4·8
Their thickness	2·5	2·3
Projection of the hinder border of the orbit from the skull . . .	2·2	2·1
Greatest breadth of the cranium between the edges of the orbits . .	9·4	8·7
Transverse diameter of the occiput near the bases of the horns . . .	5·7	5·2
Transverse diameter at the basal edges of squamosals	7·1	6·7
Vertical diameter of occiput from occipital foramen	5·0	4·3
Transverse diameter of nose at its base (widest part of maxillaries near the malar suture)	5·3	5·4
Breadth of palate between the molars (first pair of true molars) . .	2·9	3·1
Ditto ditto foremost pro-molars	—	2·3
Ditto ditto hindmost molars	—	3·3
Length of the line of molar alveoli	5·1	5·4

The measurements which relate to the horn-cores cannot be accurately compared in the fossil and recent state, as these exostoses are very fragile, and are soon reduced in dimensions when no longer defended by their cases. It will be observed that the fossil seems to have had a wider and a higher occiput than the recent bull, and these dimensions will be augmented by six per cent. if Pallas, instead of the London foot, used the Paris one, as Cuvier states that he did.

OVIBOS MAXIMUS.

A species resembling the musk-ox in the form of its second vertebra, but of which no other certain remains have been detected.

No. $\frac{90}{2}$ is a *dentata, axis,* or *peristropheus,* as the second cervical has been named by different authors. It has lost the neural spine and the parapophyses, but is otherwise very sound and heavy, few signs of decay being perceptible.

From its peculiarly massive, compact structure, it has more the aspect of the axis of a pachyderm than of a ruminant, as was remarked by an eminent comparative anatomist to whom I showed it; and it is very unlike the comparatively long, slightly made, and light dentata of the domestic ox. Its centrum and odontoid process are however constructed on the same plan with that of the musk-ox, though presenting some distinctive characters, and it is of a much greater size. The upper figures in Plate XI. represent the fossil in its present mutilated state in three different points of view.

E

	FOSSIL.	MUSK-BULL.	LARGE OX.
Length of the centrum on its sternal side, from the proximal edge of the odontoid to the distal articulating surface	4·2 in.	3·4 in.	6·1 in.
Length of centrum alone on its sternal aspect, excluding all the proximal articulation	2·8	2·2	5·0
Transverse diameter of antarticulating surface of the centrum . .	4·6	4·3	4·4
Distance from the edge of the centrum on its sternal side to the nearest side of the neural canal, measured on the mesial plane by callipers	2·2	1·9	1·5
Sterno-dorsal diameter of the distal articulating cup of the centrum .	2·5	1·9	2·6
Transverse diameter of the same cup	3·6	2·2	1·9
Transverse diameter of the centrum immediately above the parapophysis	4·3	3·6	2·3
Distance from the outside of one prozygapophysis to that of the other	3·7	2·5	3·1
Distance from the sternal front of the centrum, at its distal end, to the dorsal surface of the zygapophysis	4·5	3·6	4·2
Height of the distal end of the spine from the neural canal . .	—	3·1	2·6

The odontoid process is shaped like that of the musk-bull, but is thicker in its substance, with a more obtuse uneven edge, and not curved back so far dorsad on the sides of the neural canal. The articulating surface of the centrum falls off more convexly in every direction from the odontoid process than in the domestic ox, in which it stands out flatly and at right angles with the process, though even in that species there is a convexity of the surface towards its lateral edges. The amount of convexity in the fossil is much the same as in the musk-bull, but from some differences in the sectional outlines of the two, the articulating surface of the fossil is divided at the margin into four lobes, while in the living species there is merely a lateral rounded lobe or division on each side and a rough surface in the middle, or sternad, against which the hypapophysial knob abuts, when the head nods as in Plate V. fig. 1. The marginal lobes of the fossil are shown in Plate XI. fig. 4, and may be compared with those of the dentata of the existing species, Plate V. fig. 5. It is most probable that the hypapophysial knob of the atlas of the fossil species differs from that of the musk-ox, since instead of resting against one mesial projection it seems to have played against the middle pair of lobes, there being a re-entering notch between them where there is a projection in the recent dentata. The lateral lobes are also much chamfered away at the edge, indicating considerable lateral motion of a heavy head.

There is no vestige of any hypapophysial ridge, but the mesial part of the border of the distal articulating cup curves sacrad and is rough, the same part in the musk-ox being a smooth protuberance, but having the same general form and direction. This process in the domestic ox is a narrow projecting point of the autogenous distal articulating cup, and forms the termination of a high, thin, acute hypapophysial ridge, with which however it does not coalesce till the animal is above four years old. It is fully coalesced in the young musk-bull, which we suppose to be about that age, so as to show merely a very faint indication of its limits, while in the other cervicals of the same individual the articulating epiphyses of the centra exhibited very apparent lines of union, and separated readily in drying the bones. The distal articulating cup of the fossil is proportionally wider than in the living species, but is equally or even more concave.

The neural canal is almost round as in the musk-ox, and scarcely admits the middle finger.

The zygapophyses occupy more space than in the domestic ox, and consequently much more than in the musk-bull, and they are more stoutly made than in either. The firmness of the joint is also augmented by their articulating surfaces approaching much nearer to the centrum than in either of the other two, and these surfaces are concave, though with a slight longitudinal rise in the middle; they are convex with a shallow undulation in the musk-bull, and in the domestic ox they have a middle convexity, bounded by a concave surface on each, equal to the rise in width.

The broken roots of the parapophyses show that these processes were roundish, as in the musk-ox, and very different from the thin wing-like appendages which they form in the domestic ox, or even in the aurochs or mustush. There is little incurvature of the sides of the centrum atlantad of their roots, and the fossil consequently does not show the bold shoulder which the dentata of the musk-bull exhibits when viewed in front, as in Plate V. fig. 1, which compare with Plate XI. fig. 3.

The *dentatæ* of the aurochs and mustush, which are figured in Plate VIII. fig. 9, 10, and fig. 11, 12, reduced to half their linear size, are constructed unlike the fossil, and more in accordance with the plan of the dentata of the domestic ox. In the mustush the proximal articular surface, like that of the domestic ox, is more flat and at right angles with the odontoid (fig. 11, 12) than in the aurochs (fig. 9, 10), where it has considerable convexity both transversely and sterno-dorsad, and rounds off less abruptly into the odontoid, rendering that process stronger at the root, in accordance with the heavier horn which the aurochs carries.

The articulating surfaces of the zygapophyses differ in being longer, elliptical and acute atlantad in the mustush, while in the aurochs they are subrotund. Transversely in the mustush each of these surfaces presents a concave and a convex curve in equal proportions, and running evenly into each other; but in the aurochs there is a mesial convexity, bounded on each side by a concave curve. Neither resemble the undulating, but on the whole flattish or rather concave, zygapophysial articulation of the fossil.

The neural spine of the aurochs is stouter than in the mustush, and its knob is larger and of a different form (fig. 10 and 12). The spine itself rises more atlantad in the mustush, producing a narrower and deeper notch between it and the odontoid when seen in profile. In the musk-bull the knob is placed altogether at the other end of the spine. This was perhaps the case in the fossil, as the fractured surface of the root of the spine becomes slightly wider atlantad, as it would do in the musk-bull if the spine were broken off at the same place, while in the domestic ox the distal end of the spine is by much the thickest one, even at its root.

The parapophyses are nearly alike in the two; they are not roundish and moderately tapering as in the musk-bull, with a nearly round base; nor are they so thin and so incurved on the edges as in the domestic ox, but of an intermediate form. In the aurochs they are somewhat stouter, and stand out more laterally than in the mustush. The hypapophysial crest differs little in the two, but it ends sacrad in a thicker knob in the aurochs than in the mustush.

	AUROCHS BULL.	MUSTUSH COW.
Length of the centrum on its sternal side, from the proximal edge of the odontoid to the distal side of the articulating surface	5·0 in.	4·5 in.
Length of centrum on its sternal aspect, excluding all the proximal articulation	4·2	3·4
Transverse diameter of antarticulating surface	4·4	4·5
Height of neural spine above the neural canal, sacral end	3·2	3·1

A dentata in the British Museum, numbered 21,323, and found at Grays on the Thames, has a strong hypapophysial ridge, a proximal articulating surface 5·7 inches wide, and a length on the sternal side of the centrum, including the odontoid, of 5·3 inches,—the latter process measuring separately 1·3 inch. This I suppose to be the dentata of the *Bison priscus*.

If, as we have reason to conclude from analogy, the centra of the other vertebræ of *Ovibos maximus* are constructed as much like those of the musk-ox as the dentata No. $\frac{90}{2}$ is, no other vertebræ of the same species have been collected, unless the cervical figured by Dr. Buckland (pl. iii. f. 17) be a part of its skeleton. This specimen could not be found in the British Museum, and perhaps was not deposited there.

Until further explorations of the ice-cliffs shall have accumulated a greater mass of material, the true arrangement of the bones of the extremities in the collection must remain a matter of no small doubt.

The Indian reports of a large bovine animal inhabiting the barren grounds of Russian America, within the Arctic Circle, may refer either to the musk-ox of the more eastern districts or to another and perhaps larger species. The accounts received from the natives only are very vague.

———————————

Baron Cuvier has acknowledged the great difficulty of discriminating the species of bovine animals by detached fragments of their skeletons, and trusts chiefly to the skull for distinctive characters. Of the fossil *Bovidæ* known in his time he points out three well-marked kinds.

1st. One analogous to the musk-ox, but with which he was made acquainted chiefly by the figures and descriptions of the Russian naturalists, referred to in the preceding article. The data in his possession did not enable him to decide positively whether the fossil and the living species were identical, and a greater accumulation of fossil materials is needed to determine the question. The description of the recent skeleton of a young musk-bull which is subjoined, and the figures of the skeleton and of several skulls and other bones, will facilitate future comparisons. Up to this day a skeleton of the musk-bull has been a desideratum to comparative anatomists, which is now supplied in the Museum at Haslar; and through the kindness of Mr. Rae, of the Hudson's Bay Company, I hope to receive in a few months another skeleton of this animal, and also several of some of the other large northern ruminants.

2nd. The next kind is mentioned by Cuvier as having been an ox of greatly superior size to the domestic cattle of France, and as having its very large horns differently curved and inclined. This animal he thinks was not extirpated until the time of the first dynasty of French kings ; and he quotes a passage from Gregory of Tours which relates that King Gontram put several of his officers to death because, in breach of the game-laws, they had killed an *urus* (erroneously called a "buffalo" by the historian) in one of the royal forests, situated among the mountains of the Vosges*. He even thinks that the "thur," which Herberstein mentions and figures in his 'Commentarii de Rebus Muscoviticis' (p. 82) as existing in his time (1520–26) in certain parts of Mussovia, on the borders of Lithuania, was in fact the *urus*, and very distinct from the "aurochs" or "zubr" of the Poles, which Herberstein also figures† (*Op. cit.*, p. 143).

* Grégoire de Tours, lib. x. cap. x. (Oss. Foss.) "Gregorius Florentinus, geboren An. 544, wurde 573 bischof " (Meyer).

† These figures are copied by Gesner. (Oss. Foss.) " Uros sola Massowia, Lithuaniæ contermina,

The Romans first became acquainted with this wild and fierce beast when they carried their arms into northern Gaul and Germany; and Cæsar describes the *urus* as nearly equal to the elephant in size, with the aspect, colour, and shape of the bull, but differing from the domestic animal in the spread, shape, and appearance of its horns. It possesses, he says, great strength and speed, and does not hesitate to attack either man or wild beast. Vessels formed of its horns, and edged with silver, were used in the most sumptuous feasts*. Seneca and Pliny give the characteristic differences between the *urus* and *bison*†. In Cæsar's time the urus was considered as peculiar to the forest of Hercynia (the Black Forest); and it does not appear ever to have inhabited the great woodland districts of Russia, nor within the European historical period to have frequented the coasts of the German Ocean or British Channel. But in the German 'Niebelungen Lied' of the twelfth century mention is made of four stark "*Ure*" as having been slain in a great hunt held in the Forest of Worms. In pre-historical periods it seems to have been more abundant in Europe, and more widely spread, since its remains (or at least the bones of a great ox not hitherto distinguished from it) have been found in various parts of Great Britain, in Germany, France, and in the north of Italy, generally in the more superficial deposits; but Professor Owen has adduced instances of their association with the bones of the mammoth, rhinoceros, bear, boar, and horse, in several parts of England, such as Grays on the Thames, Clacton in Essex, and Molksham on the Avon, in drift sand or mud overlying the London clay.

In 1825 Bojanus gave the figure of an entire skeleton of the fossil ox, contrasting it with that of the aurochs, and bestowing upon it the scientific designation of *Bos primigenius*. Meyer in 1832 published a summary of what was then known of the fossil *Bovidæ*; and in 1846 Professor Owen, in his excellent 'History of British Fossil Mammals,' described the remains of the "great fossil ox" found in the British Islands, adding a compendium of its literary history, collected by preceding authors, but chiefly by Cuvier. He gives two excellent views of a skull found in Scotland (pp. 498, 507, fig. 208, 210), and one of a metatarsus dug up on the banks of the Thames‡.

habet; quos ibi patrio nomine '*thur*' vocant . . . non est magna horum copia; suntque certi pagi, quibus cura et custodia eorum incumbit: nec fere aliter quam in vivariis quibusdam servantur."—*Herberst. de Rebus Muscoviticis*, p. 81. (Oss. Foss.) M. Goldfuss states that the domestic bull is called "Ur" to this day in some Swiss cantons, and this is evidently the origin of the name adopted by the canton of Uri.

* Cæsar, De Bello Gallico.

† " Tibi dant variæ pectora tigres,
Tibi villosi terga bisontes,
Latisque feri cornibus uri."—*Senec. Hippol. Act.* i. 63. (Oss. Foss.)

" Jubatos bisontes, excellentique vi et velocitate uros, quibus imperitum vulgos bubalorum nomen imponit."—*Pliny, Nat. Hist.* lib. viii. c. xv. (Oss. Foss.)

‡ *De* Uro nostrate *ejusque sceleto Commentatio, scripsit, et* Bovis primigenii *sceleto auxit*, Lud. H. Bojanus, 1825. (Nov. Act. Med. Phys. Acad. Cæsar. Leop. Carol. Naturæ Curiosor. tom. xiii. pars ii. p. 414.)

Ueber fossile Reste von Ochsen, deren Arten und das Vorkommen derselben; von Hermann von Meyer. Dec. 1832. (Nov. Act. Phys. Med. Acad. Cæs. Leop. Carol. Nat. Curios. Vratislaviæ et Bonnæ, 1835.)

N. C. De Fremerij over eenen Hoorn, enz., van *Bos primigenius*, in ' Nieuwe Verhandlingen van het Koninglijk-Nederlandsche Instituut van Wetenschappen, etc.; te Amsterdam, 1831.'

Notice sur les Bœufs fossiles de Sibérie, par G. Fischer; publié dans le Bulletin de la Société Impériale des Naturalistes de Moscou. Seconde année, 1830. No 1, avec 6 planches.

Cuvier remarks that though the crania of the great fossil ox indicate an animal greatly superior in size to the domestic oxen of France, they do not appear to have belonged to a beast much bigger than the great oxen of Hungary, Podolia, and Sicily.

Bojanus, applying the following verses, written by Hussowezk, of Cracow, in 1523, to the bison, treats the extent of the horns as an exaggeration; but if the poet alluded to the uri spoken of by Herberstein as then existing in the parks of Poland, there seems to be nothing improbable in three men sitting between the widely-spread horns of that animal. Allowing 14 inches to each man, the room required would be but 42 inches, or nearly that assigned to the span of the horn-cores of the Athol fossil skull figured by Professor Owen. The distance between the points of the horns of a Spanish ox in Haslar Museum is 44 inches.

> " Nascitur et fieri corpore tanto solet,
>   Ut moriens, si quando caput vi victa reclinet,
>   Tres sedeant inter cornua bina viri."

The third kind of fossil bovine species is briefly characterized by Cuvier as differing almost in nothing from the aurochs, or the zubr of the Poles (*Bison Europæus*), and of which Professor Owen, in 1846, states that no satisfactory specific distinction has been detected in the fossils when compared with the bones of the Lithuanian aurochs, though the remains of the ancient bisons attest their larger size, and longer and somewhat less bent horns. He adopts, however, Bojanus's specific name of *priscus* for the fossil. The Romans on invading Germany found the bison a more general inhabitant of the forests than the urus, and it was known previously to Aristotle under the names of *Bonasus*, *Bolinthus*, *Monepus*, and *Monapus*, as a native of that part of Thrace which is now named Bulgaria (Cuv. Oss. Foss., vol. iv. p. 112). In the present day the aurochs survives under the protection of strict game-laws in Lithuania, not very remote from its ancient haunts in Pæonian Thrace; but it is not found alive in the boundless forests of Russia or northern Asia.

To the 'Ossemens Fossiles' I must refer for the notices of the bison by Aristotle, Seneca, Pliny, Oppian, Pausanias, and other ancient writers quoted by Cuvier; but I may add one which has escaped him, and which has relation to the comparative size of the domestic cattle of the dwellers on the Lower Rhine, when compared with the uri of the same quarters.

" Eodem anno (A.D. 15) Frisii, transrhenanus populus, pacem exuêre, nostrâ magis avaritiâ, quam obsequii impatientes. Tributum iis Drusus jusserat modicum, pro angustiâ rerum, ut in usus militares coria boum penderent, non intenta cujusquam cura quæ firmitudo, quæ mensura; donec Olennius e primipilaribus, regendis Frisiis impositus, terga *urorum* delegit, quorum ad formam acciperentur. Id aliis quoque nationibus arduum, apud Germanos difficilius tolerabatur, queis ingentium belluarum feraces saltus, modica domi armenta sunt. Ac primo boves ipsos, mox agros, postremo corpora conjugium ac liberorum servitio tradebant. Hinc ira et questio," etc. (Tacitus, Ann. iv. § 72. A. U. 781.)

It is probable that in this passage Tacitus speaks of the wild bison rather than of the urus, which Cæsar had said was confined to Hercynia. Bojanus, commenting upon the size which the aurochs of the present day attains, states that they are not observed to grow higher than 5 feet (*pieds du roi*) and scarcely to reach 7½ feet in length, and never to surpass that magnitude, nor do any but the aged males reach such a size. Formerly they grew to greater dimensions, if no mistake occurs in the measurements adduced in the following instances, collected by Bojanus. Count George Frederick, in 1595, killed one near Friedrichsburg 7 feet high, 13 feet long, and weighing 995 lbs. As it is possible that some uri were still alive at that date, this record, even if perfectly true, is not decisive as to the occasional size of the aurochs; but the two following, from the lateness of the periods, could have referred only to the aurochs. Augustus III. killed an aurochs in Poland, in 1752, that weighed 1450 lbs., and John Sigismund killed one in 1612 that weighed 1770 lbs.

With respect to the fossil skulls having convex frontals and horns originating anteriorly to the occipital crest, as in the aurochs, of which a considerable number were examined by Baron Cuvier, that acute and accurate naturalist refers them all to one species, which is found fossil throughout the whole northern part of the two continents, in Italy, Germany, France, Prussia, England, Siberia, and North America. In the first edition of his ' Ossemens Fossiles' he considered these fossil skulls from Europe as belonging to the aurochs of the present day; but coming afterwards to discover that they did not resemble the skull of the living aurochs more closely than that of the latter resembles the skull of the American bison, he saw no reason for not attributing the large fossil skulls of Europe, Asia, and America, to a third species, distinct from both the aurochs and the (*mustush*) American bison*.

Bojanus, in 1825, bestowed the name of *Bos* (*Bison*) *priscus* on the fossil species, and in 1830 M. G. Fischer named the Siberian skulls resembling that of the aurochs *Bos latifrons*, characterizing them as being broad between the horns, and also elongated and flattened, with prominent orbits, broad maxillæ, and broad and oval palato. (Fischer, op. cit. antè.)

Cuvier had not (in 1825) ascertained the geological age of the deposits in which the bones of the *Bison priscus* are found imbedded; but Meyer, writing ten years later, states that they are found in the true diluvium spread over the northern half of the earth, in Europe, Asia, and America, associated for the most part with *Elephas primigenius*, *Rhinoceros tichorhinus*, or *Hippopotamus major*, *Equus fossilis*, *Cervus eurycerus*, and more rarely with *Bos primigenius*. *Bos* (*Ovibos*) *Pallasii*, he adds, is found in Siberia and North America, in which last country the *Bos* (*Bison*) *bombifrons* also occurs. (Meyer, op. suprà cit.)

From Professor Owen's ' History of British Fossil Mammals' we collect that skulls of the *Bison priscus* have been dug up at Woolwich from a stratum of dark-coloured clay lying beneath layers of brick-earth and gravel, thirty feet below the surface; also in the " fresh-water newer pliocene deposits" at Walton, in Essex; out of a brick-field at Ilford, in the same county; and from " fresh-water drift" in Worcestershire. In the brick-earth of Woolwich and Ilford, which underlies a layer of sand, with pebbles and concretions containing shells of *Unio* and *Cyclas*, the remains of the mammoth and rhinoceros are unquestionably associated with those of the *Bison priscus*. At Beilbecks, in

---

* " J'ai tout lieu de croire que tous ces crânes fossiles à front bombé viennent d'une seule et même espèce" (p. 146). " L'espèce est donc enfouie en réalité dans toute la partie boréale des deux continens, puisqu'on en a d'Allemagne, d'Italie, de Prusse, de la Sibérie occidentale et orientale, et de l'Amérique. Dans ma première édition j'avais considéré les crânes fossiles d'Europe comme appartenant à l'aurochs ordinaire, et ceux de Sibérie comme provenant d'une espèce perdue; maintenant que j'ai reconnu les uns et les autres pour être de la même espèce, il s'agirait de savoir s'ils seraient tous de l'aurochs; mais comme je viens de constater aussi qu'ils ne ressemblent pas plus à l'aurochs que celui-ci ne ressemble au bison d'Amérique, et comme ces deux animaux sont distincts par l'espèce, on ne voit pas pourquoi celui qui a produit les grands crânes fossiles ne serait pas d'une troisième espèce, aussi distincte que les deux premières, et dont les caractères auraient tenu à d'autres parties qu'à la tête. La grandeur de ses cornes pourrait déjà le faire soupçonner, car les plus vieux bisons et les plus vieux aurochs n'ont que des cornes médiocres. M. Hacquet m'écrit que les plus grands individus n'ont pas des noyaux de cornes de plus d'un pied de long." (Cuv. Oss. Foss. An. 1825, p. 148, troisième édition.)

Yorkshire, also, the skull, with cases of the horns and the teeth of the great fossil aurochs, accompanied by land and fresh-water shells, are found along with remains of the *mammoth, rhinoceros, Felis, large horse, large deer, wolf,* etc.   (Owen, lib. cit. p. 496.)

After an inspection of the skeleton of an adult male aurochs presented by the Emperor of Russia to the British Museum, Professor Owen adds to what he had said in the preceding pages of his work the following passage :—" The skull shows the same expanse, convexity, and shortness of the frontal region, and the same angle between this and the occipital region, as does the fossil skull of the *Bison priscus;* the horn-cores have the same advanced origin and the same direction; these, however, are relatively shorter than in most of the fossil skulls, and the general size of the existing aurochs is less than that of the ancient or fossil specimens.   Admitting with Cuvier that such characters are neither constant nor proper for the distinction of species, one may recognize in the confined sphere of existence to which the aurochs has been progressively reduced precisely the conditions calculated to produce a general loss of size and strength, and a special diminution of the weapons of offence and defence.   I cannot perceive, therefore, any adequate ground for abandoning the conclusion to which I had arrived from a study of the less perfect materials available to that end, before the arrival of the skeleton of the Lithuanian aurochs, viz., that this species was contemporary with the mammoth, the tichorhine rhinoceros, and other extinct mammals of the pliocene period." (Owen, lib. cit. p. 515.)

With respect to detached bones of the fossil *Bovidæ,* the slenderest of the long ones are usually referred to the *Bison priscus,* and the thick ones to the *Bos primigenius.*   It may be a question, however, whether the fossil species are not more numerous than the caution of the able palæontologists to whom we are indebted for what has hitherto been published on the subject is willing to admit.   This notion is put forth, however, rather as a suggestion intended to promote further investigation than as an opinion supported by proofs, since, from my residence at a distance from London, and other circumstances alluded to in a preceding page, I have not had an opportunity of mastering the subject by a satisfactory study of the skulls and detached bones of the skeleton of the European fossil bisons, and have merely noticed that several large vertebræ found at Grays, and in other deposits, seemed, from the cursory inspection I was able to bestow upon them in the British Museum, to exhibit different characters from the corresponding bones of the common ox, the Lithuanian aurochs, American bison, or the bovine vertebræ dug out of the ice-cliffs of Eschscholtz Bay.   On this brief and imperfect comparison, therefore, I would found no argument; but after long consideration of that part of Captain Kellett's collection of bovine fossils from the ice-cliffs which is now in Haslar Museum, I have convinced myself that among these there are the remains of one, and perhaps two, species of the bison type, related as closely to the American bison as to the aurochs, and one of them differing from both in the thickness of the long bones of the extremities, which bear in that respect a greater similarity to the bones of *Bos primigenius.*   None of the skulls of the great fossil ox have been detected in the ice-cliffs, nor as far as I have learnt in any part of Siberia, and both the vertebræ and the cannon bones brought from the cliffs differ from those of the domestic ox, which are said to be exactly similar, except in size, to those of the great fossil species.   Professor Owen records the discovery of nearly an entire skeleton of the *Bos primigenius* at Clacton, by Mr. Brown, of Stanway, but I have not been able to ascertain where that very important relic of a past age is deposited.   Even a short inspection of it would have determined many of the doubts which have perplexed me in this investigation.   The two species which I suppose to have found a tomb in the ice-cliffs are independent of a fossil musk-cow of the size of the existing species, and of the *Ovibos maximus,* which resembles it at least in the peculiar form of the second cervical, and which is described in page 26, and figured in Plate XI. (fig. 2, 3, 4), and a fragment of the sixth cervical of perhaps the same species.

It belongs to this inquiry to notice that the bone deposit of Bigbone Lick, on the banks of the Ohio, has yielded two skulls of different species of the bison type: one found by Mr. Peale, figured by Cuvier in the 'Ossemens Fossiles,' pl. xii. f. 2, and referred by him to the fossil aurochs, and the other described by Caspar Wistar, under the name of *Bos* (*Bison*) *bombifrons*, in the Transactions of the Academy of Philadelphia, and which presents very peculiar characters.

The living *Bison Americanus*, which for convenience we shall generally designate by its native Cree name of "mustush," is properly an inhabitant of the prairies, and does not range further north than the Horn Mountain, in latitude 62°. Until of late years it was unknown to the westward of the Rocky Mountains, but since it has been so much persecuted on the prairies some herds have found their way westward into the valleys traversed by the southern tributaries of the Oregon. The large herds keep strictly to the open prairies, but small bands penetrate the woodlands, and the individuals of the forest mustush are generally of a large size.

BISON PRISCUS? (Bison latifrons, *Fischer?*) *Fossil Bison of Arctic America.*

The reference of the following bones to this species is made with doubt, since, as has been already mentioned, circumstances have prevented a due comparison of them with recognized examples of the great fossil aurochs of Europe. Though several of the skulls from Eschscholtz Bay bear a strong resemblance to Meyer's figures of the fossil aurochs, a doubt of their specific identity arises from Professor Owen having seen no distinctive characters between the living and fossil aurochs of Europe, while in the fragments of crania and detached bones from the ice-cliffs the resemblance is in many respects as great, if not greater, to the same parts of the mustush. One character of the ice-cliff fossils that may be readily seized is the lateral prolongation of the paroccipital condyles into a shallow concave trochlea, similar to that of the mustush, and which in the latter has reference to an articular surface of the atlas, formed on the broad edge of a lateral notch: while in the living aurochs the brim of the antarticular cup of the atlas presents at the same part a rather acute edge, rounded in outline. The trochlea of the fossil is represented in Plate VI. fig. 6; that of the mustush in fig. 4; and may be contrasted with the condyles of the aurochs (fig. 2), in which the lateral projection is wanting. It may be necessary to remark that the apparent notch between the articular surfaces of the prozygapophysis and centrum in fig. 1 is owing to the obliquity of the point of view of the antarticular cup. The shallowness of the notch, which scarcely breaks the continuity of the brim, is shown by the side view in fig. 2, and the thinness of its walls by the bird's-eye view in fig. 3. The width of the notch in the mustush, and the broadness of its walls, lined on the edge with articulating cartilage, and the consequent wavy outline of the brim of the antarticular cup, is shown in fig. 5, 6, and 7 of the same plate, in which all the figures are reduced to half the linear size. The atlas of the musk-bull shows this kind of conformation, still more developed, in accordance with the more complete trochleas of the condyles. See Plate V. fig. 1, 3.

Comparisons have been made between the following fossil bones and the adult male aurochs skeleton in the British Museum, the skeleton of a mustush cow and the skull of a mustush bull in the same collection; also with the skulls and some detached bones of large Spanish and Lincoln-

F

shire oxen in the Haslar Museum, and with the complete skeleton of an aged Alderney cow in the same institution.

No. 24,589. This number is affixed to a fragment of a *skull* belonging to Captain Kellett's first collection, now in the British Museum. It has lost all the bones of the face and jaws. The upper figure (fig. 1) in Plate VII. gives a front view of it contrasted with similar views of the same portion of the skulls of adult males of the aurochs and mustush, all of half the linear size.

Bojanus states that the horns of the aurochs are round throughout their entire length, and nowhere compressed, and the horn-cores of this fossil appear to be nearly so to the eye; but when measured by callipers they are found to be somewhat depressed at the base, their vertical diameter bearing a proportion to their antero-posterior one of 30 to 32. They are less curved than those of the aurochs, and are directed mainly outwards, with a slight inclination towards the plane of the occiput; their curve is a segment of a circle of 28 inches radius, and has the same aspect with the frontal. That side of the core is smoother and the others are deeply and angularly grooved in the direction of their length, with little of the spiral twist exhibited by the grooves of the aurochs horn-cores.

Though the horn-cores of the fossil are longer, and the occipital plate wider and higher than that of the adult male aurochs, the angle of the temporal fossa is much narrower, owing to the more iniad (or backward) direction of the horns. When the points of the horn-cores of No. 24,589 are brought into the same line of view, so that one of them hides the other, the occipital ridge is concealed by the posterior border of the cores; but in the skull of an aurochs similarly placed the angle of the temporal fossa is exposed. The angle which the plane of the occiput makes with the frontal is somewhat more open than in the aurochs, owing chiefly to the occipital plate near the foramen magnum and the condyles sloping more in the direction of the spine, or sacrad. The occipital ridge is much alike in the two, and the rough surface which represents the spine is unlike that of the mustush (Plate VI. fig. 3), and more like the aurochs (Plate VI. fig. 1), only it is a little more prominent in the fossil. A difference in form of the condyles has already been mentioned, and may be seen in Plate VI., where fig. 1 and 2 give upper and under views of the condyles of the aurochs, and fig. 5, 6, of the fossil species, though not of this individual specimen. The condyles of the fossil are also longer, more oval, and more oblique than those of the recent aurochs.

The basi-occipital is not so wide as that of the adult bull aurochs, is less flat, and has a smooth rounded mesial groove in the direction of its length, formed by the greater elevation of the shoulders, which abut against the confluent notches of the centrum of the atlas, and by the rounded elevations of the edges of the bone continued from these shoulders to the protuberances at the junction of the basi-sphenoid. The basi-sphenoid has a less distinct hypapophysial ridge than the living aurochs, but it can be traced.

The frontal of the fossil skull is decidedly flatter, both transversely and longitudinally, than in the aurochs; a straight ruler laid across between the horns rests on the swelling-up of the frontal, to unite with the horn-cores, the middle part being free, but one similarly placed in the aurochs rests with its middle third on the convexity of the frontal, the ends being free. On comparing the fossil skull with one of an adult male mustush (*Bison Americanus*), the grooving of the horn-cores seemed alike in both; but the cores of the latter were more directed backwards, and had a more decided upward curve, while at the same time they were considerably shorter. The occipital foramen of the mustush bull is conspicuously smaller. The basi-sphenoid is also smaller in this animal than in the fossil, and much more so than in the aurochs; but the basi-occipital of the mustush, though also more slender, otherwise resembles that bone in the fossil. The occipital plate of the fossil exceeds that of the mustush in height, and in a still greater proportion in width at the most lateral parts.

Distance from the mesial plane of the occiput to the tip of one horn-core . . 16·5 inches.
Spread of the horn-cores from tip to tip in a straight line . . . . . . 30·0
Vertical diameter of base of horn-core . . . . 3·0
Antero-posterior ditto . . . . 3·7
Distance from the rough ring at the base of one horn-core to that of the other across
    the forehead . . . . 12·0
Distance from the outside of the accessory trochlea of one condyle to that of the other 6·3
Height of the occiput from lower edge of the foramen magnum to the upper edge of the
    occipital ridge on the mesial line . . . . 6·2
Width of occiput at its greatest breadth or between the reverted edges of the squamosals 11·3
Span of the horn-cores in the aurochs bull (Plate VII. fig. 2) . . . 24·5
Ditto   ditto   mustush bull (Plate VII. fig. 4) . . . 22·0

There is a second skull in Captain Kellett's collection in the British Museum, which is split into two parts, but is otherwise more perfect in many places than the preceding one. It bears also the number 24,589.

No. 105. This number refers to the *left horn*, with a small portion of the frontal, now in Haslar Museum, being part of Captain Kellett's second collection. When brought home the core was covered by its horny sheath, and the latter still retained its dark blackish-brown colour, and much elasticity in its plates, but portions had scaled off towards the tip, and in the basal half the successive laminæ, being worn so as to present a tiled appearance, formed about a dozen large circular rugæ, each of them splitting into ten or twelve laminæ. The horn case exceeds the length of the core by about 4 or 5 inches, and has the same curvature. The grooves of the horn-cores have corresponding folds in the sheaths, which are visible in all the interior layers, and can be traced even in the superficial one.

The core was separated from its case by boiling, and a side view of it is given in Plate XIII. fig. 3, as seen from its fascial side. It must be observed, however, that the figure has been reversed in printing from the stone drawing. This core has a single curve without any spiral twist. On its coronal aspect, which is its concave side, the large grooves are shallow and exist only near the point; eight or more large ones occupy the rest of the circumference, extending from near the base to near the point. Their edges are acute, and they are all more subdivided in portions of their length by acute ridges. The whole surface of the core is also rough and coarsely cellular. On the basilar aspect, where the cellular core coalesces with the smooth frontal, the surface is flattened or concave, producing a neck, and a smooth groove (not visible in the figure) traverses about one-third of the circumference.

Length of the horn-core from its basal ring, measured along the curve of its con-
    cave side . . . . 10·5 inches.
Its vertical diameter at the thickest part of its base . . . 2·6
Its antero-posterior diameter there . . . . 2·8
Circumference at the base . . . . . 8·4

No. 106 is the horny case without the core, similar to the case of No. 105, but with the layers more detached and changed in colour, being pale and partially stained reddish or orange-brown. The longitudinal grooves and folds appear strongly in all the laminæ except near the points of the bone, where they become obsolete.

Plate iii. f. 2, in the Appendix to Beechey's Voyage, seems to represent a horn of this kind, but no grooves are shown. As Dr. Buckland (p. 336) mentions that specimen as being in a fresh state, and probably in his opinion not fossil, the external layers of the horn were doubtless entire, and covered therefore the grooves of the interior ones, supposing it to have been of this species.

No. 104 is a *horn-core* also of this species, which is much worn, having been long exposed to friction among the gravel. It has lost much of its animal matter. No. 105, on the contrary, is heavy in comparison, and the frontal fragment adhering to it is hard like a recent bone with smooth surfaces, but it is of a dark colour, and tried by the tongue shows some absorbency.

No. 140. This is a small fragment only of a skull, and consists of the basi- and paroccipitals, with their surfaces more or less mutilated, and the true form of the condyles obliterated. The width of the basi-occipital equals that of the adult aurochs bull, and exceeds that of the mustush. This bone has not, however, the flatness which it shows in the aurochs, being transversed by a smooth longitudinal groove, into which the fore finger is readily received. Its shoulders, which abut against the notches in the brim of the atlas, are much developed. The super-occipitals show several large cells separating the tables of the skull at the edge of the fractured bone, an inch above the foramen magnum. In the groove of the basi-occipital the resemblance of the fossil to the mustush is considerable. The specimen is in Haslar Museum. On comparing this fragment with the skull of a large Spanish ox, the occipital foramen appeared larger in both its dimensions, and its condyles measure more to their extreme edges.

Distance between the most lateral part of one occipital condyle and that of the other . 5·2 inches.
Extreme width of the shoulders of the basi-occipital . . . . 2·7
Breadth of the basi-occipital where it coalesces with the basi-sphenoid . . 1·9
Thickness of the substance of the bone at the latter place . . . 1·1
Vertical diameter of the occipital foramen . . . 2·0
Transverse ditto between the middles of the condyles . . . 1·5
The extreme width of the condyles in the bison *cow* is . . . 4·5

No. 24,576 is a fragment similar to the preceding one, belonging to Captain Kellett's first collection, and deposited in the British Museum. As its condyles are perfect, they have been represented in Plate VI.; fig. 6 giving a view of them on the basilar, and fig. 5 on the occipital aspect; both of the natural size.

Outside measurement of the condyles . . . . . 5·1 inches.
Transverse diameter of the occipital foramen . . . 1·6
Antero-posterior diameter of ditto . . . . 1·9
The length of the basi-occipital, from the edge of occipital foramen to the commencement of the basi-sphenoid . . . 2·0

The skeleton of a mustush cow, which is in the British Museum, has a width between the lateral parts of its condyles of 4·5 inches, but the body of its basi-occipital is as long as in the preceding fossils that have much wider condyles.

No. 24,373 is another piece of a skull, being part of Captain Kellett's first collection, and is also in the British Museum. It is a similar fragment to the last, but is not so perfect.

In the five fragments of skulls from the ice-cliffs above mentioned, the condyles are wider than in the adult male mustush in the British Museum, with which we compared them, but they strongly resemble in that part a fossil skull in the same collection which was dug up at Brentford, and which is, it is presumed, a recognized example of the European *Bison priscus*. Metatarsal and metacarpal bones from the same locality are referred by Professor Owen to that species rather than to *Bos primigenius*, from their comparative slenderness (Brit. Foss. Mam., p. 495)*.

No. 24,590. A fragment of the *alveolar processes* of the left maxilla, containing the first and second true molars. The molars are larger than those of a domestic cow, and the second one much resembles the corresponding tooth of the aurochs bull, the crown however being rather thicker transversely in the fossil, and the interior folds of enamel enclosing a wider space. Externally the columns of enamel are more prominent and more acute. The difference, however, is slight, in all these particulars.

The *skull* figured by Meyer (*Op. cit.* tab. xi. f. 10, p. 103) shows that the *Bison priscus* might have horn-cores of a great length, without any extraordinary increase of thickness, and I have therefore separated from this species the short thick heavy core (No. 91) from Eschscholtz Bay, and have moreover doubts about the propriety of some of the very thick cores referred to it by Cuvier being properly so assigned, notwithstanding the deference that one feels inclined to pay to his high authority.

No. 24,576 refers to an *atlas* in the British Museum, which forms part of Captain Kellett's first collection. Different views of it are given in Plate X., of the natural size.

Transverse diameter of antarticular cup	4·6 inches.
Sterno-neural diameter of same cup	2·2
Length of the centrum on its sternal aspect	2·2
Length of vertebra from edge of prozygapophysis to distal articulation . .	4·0
Transverse diameter of post-articulation	4·3

This fossil atlas is rather intermediate in form between the atlas of the common ox and the mustush than like that of the aurochs. It is of much less solid structure than that of the fossil atlas No. 90, p. 42. In the brim of its antarticular cup it is not unlike that of the domestic ox, but the process between the lateral shallow notch and the deeper one in which the shoulders of the basi-occipital play is rather more inflexed; the edge of the lateral notch, on the other hand, is a little everted, affording a surface on which the trochlea of the condyle can move. This produces the wavy outline of the brim shown by the direct view of the cup in fig. 4. The sternal pair of notches in the brim are divided from one another by a median furrow, a slight vestige of which only is seen in No. 90, Plate XII. fig. 1; a very distinct one exists in the musk-ox, and a less conspicuous one in the aurochs and mustush, but which is replaced by a ridge in the domestic ox. The transverse diameter of the antarticulating surface just equals that of the bison-cow skeleton in the British Museum, but the sterno-neural diameter is one quarter more, which is due principally to the greater thickness of the centrum. The notch between the prozygapophyses (fig. 2, 3) is acute and unlike that of the fossil No. 90, or any of the three living species it has been compared with. The distal articulating surface represented in fig. 5 differs little from that of the aurochs, but the lateral processes are united to it more evenly, as in atlas No. 90, and not rounded off with a twist, as in the aurochs or domestic ox.

The hypapophysial tubercle is similar to and similarly situated to that of the mustush, being more approximated to the distal articulating surface than that of the aurochs. The depressions on each side

* In referring the vertebræ and other detached bones in the collection to this or the other species, we have been guided chiefly by size—the less massy ones being ascribed to *Bison priscus*, and the more strongly framed ones to *Bison crassicornis*, care however being had to associate those which had a similar type.

of the centrum under the arterial foramina (fig. 2) are neither so deep or so abrupt as in No. 90, and are more like the same part in the atlas of the mustush than in that of the aurochs.   They run out gradually on the sides of the centrum, much as in the atlas of the domestic ox.   The neural spine is like that of the domestic cow, but is more prominent.   In the interior of the neural canal the pit on each side under the arterial foramen exists, but the connecting transverse furrow is obsolete (fig. 6).

On the whole this atlas has more the character of that of the mustush than of any other living species with which it has been contrasted.   In the flatness of the articulating surfaces of the zygapophyses it is like the mustush, while in the aurochs these surfaces are convex and more undulated. The size of this atlas does not indicate an animal of greater magnitude than the existing mustush.

No. 126, *a fifth cervical*, differs from the corresponding vertebra of the aurochs in having a roughish prominence on the back of each zygapophysis, and a deeper notch between these processes. It greatly resembles the fifth cervical of the American bison, but has bigger arterial foramina, and is of rather smaller dimensions.   It evidently belonged to an adult animal, and it differs from No. 117 (referred to *Bison crassicornis*) not only in size but in the inferior development of the tubercle on the zygapophyses.   They are constructed, however, generally on a very similar model. The metaphysial point, which rises atlantad beyond the prozygapophysial articulation, is the same in both.   The most apparent difference is the considerably greater prominence of the hypapophysial ridge in No. 117, and the consequently deeper depression between the ridge and the parapophysis; and this difference is equally perceptible when No. 126 is compared with No. 116, which is also referred to *B. crassicornis*.   The hypapophysial ridge is less acute than in the common ox.

	NO. 126.	LARGE DO-MESTIC OX.
Length of centrum on its sternal aspect, including crown of articulation	3·6 in.	4·1 in.
Ditto ditto ditto excluding the crown . . .	2·5	2·6
Ditto ditto neural side, excluding the crown	2·4	2·4
Length from proximal edge of prozygapophysis to distal edge of zygapophysis	3·7	3·6
Sterno-neural diameter between the sternal side of the articulating ball and the crown of the neural arch between the prozygapophyses . .	2·7	2·7
Lateral diameter measured in the interval between the prozygapophysis and the lateral processes on the outer sides of the arterial foramina . . .	2·8	2·9
Lateral diameter between the outer sides of the prozygapophyses .	3·7	3·7
Lateral diameter between the outer sides of the zygapophyses . . .	3·4	3·8

No. 132.   A fragment of the *seventh, or eighth, or ninth dorsal*, consisting of the centrum, neural canal, and parts of the processes.   Some of the surface is still perfect, but much of the animal matter has decayed.

In the domestic cow the eighth, ninth, and tenth are the most compressed of the dorsals, and have the sharpest hypapophysial ridges, and the seventh has the shortest centrum.   The fossil corresponds in length of centrum to the ninth dorsal of the American bison-cow, and agrees with it in other respects as far as the characters can be made out.   It scarcely exceeds in size the ninth dorsal of an old English cow, and is as much compressed, with as sharp a hypapophysial ridge.

	NO. 132.	NO. 131.
Length of centrum, including convex antarticulating surface	2·5 in.	2·9 in.
Ditto ditto excluding the articulating surfaces	2·4	2·5
Lateral diameter between the outer edges of the proximal articulations for the head of the ribs	1·9	2·3

NO. 132. NO. 131.

Lateral diameter at the outer edges of the distal articulations for the heads of
the ribs . . . . 2·1 in. 2·5 in.

No. 131. This *dorsal* has more of the diapophyses remaining than the preceding, but it is otherwise much in the same state of decay. It is bigger than any of the dorsals of the American bison-
cow, from the sixth to the ninth, but is very similar to the eighth in form.

No. 134 is most probably the *third lumbar* from the sacrum. It is considerably bigger than
the corresponding vertebra of the adult aurochs or mustush, but much smaller than lumbar No. 123,
referred below to *Bison crassicornis*. In form it approaches most nearly to that of the mustush, but
differs in the shape of the prozygapophysial tubercle, and this part is more evenly rounded and less
flattened laterally than in No. 123 (*Bison crassicornis*). The depressions on each side of the neural
spine are also different in these two, being deeper in the smaller vertebra.

Length of centrum, sternal side, including convex antarticulating surface . .	2·8 inches.
Ditto ditto ditto excluding ditto	2·3
Sterno-neural diameter of centrum, proximal end . .	1·7
Length of centrum, neural side . . .	2·6
Length from proximal edge of prozygapophysis to distal edge of zygapophysis .	3·8
Lateral diameter of centrum at the notch behind the roots of the diapophyses .	1·8
Sterno-neural diameter to crown of neural arch, distal end	2·7
Lateral diameter between the outsides of the metapophyses (about) .	2·5
Ditto ditto between the outsides of the zygapophyses . .	1·7

No. 87. Part of the *right scapula*. The glenoid cavity has exactly the form of that of the
mustush, though it is larger than that of the female skeleton in the British Museum. It fits the
humerus No. 81. In general outline the glenoid cavity is obtusely ovate, with a very shallow, wide,
concave notch on the lateral or dorsal margin, extending to rather more than a fourth of the whole
circumference. It is very different in shape from the glenoid cavity of the musk-cow, which is more
elliptical, and not nearly so obtusely rounded on the ulnar side of the margin; the curved deficiency
of the lateral border is more extensive and more decided when viewed laterally, but not in looking
directly at the glenoid cavity, and the notch on the sternal side, close to the base of the coracoid
tubercle, which is so distinct in the musk-cow, does not show at all in the fossil nor in the domestic
cow. The coracoid tubercle has a different shape from that of the musk-bull or domestic cow, and
the neck of the scapula appears to be broader in proportion than in the latter. The spine at its
origin is as near the atlantal border of the scapula as in the domestic cow, and nearer than in the
musk-cow.

Longest diameter of glenoid cavity	3·0 inches.
Greatest transverse diameter which occurs in the third nearest the sacral or ulnar border	2·4
Width of the neck dorsad of the coracoid tubercle	3·3

No. 88 is a fragment of a *right scapula*, more decayed than the preceding, and with the borders
of the glenoid cavity partially worn, so that its true form cannot be ascertained. It is of equal dimensions with the preceding one.

No. 81. A *right humerus*, defective in the edges of the proximal joint and base of the tubercles,
and in the back of the condyles; the surface of the rest of the bone is in excellent preservation, and

there has been little loss of animal substance. It differs in no striking character from the humerus of the domestic ox, and though a considerably stouter bone than the humerus of the adult musk-cow, particularly at the condyles, it is only three-quarters of an inch longer: it is almost half an inch shorter than the humerus of the young musk-bull. It is larger than the humerus of an old Alderney cow, and is three-quarters of an inch wider at the condyles. A full-sized Lincolnshire ox is about a quarter of an inch wider at the elbow-joint than the fossil. It is decidedly different in form from the humerus of the musk-bull, and there are characters in the articular surfaces of the condyles by which it may be distinguished from the humerus of a domestic ox. I have not made the same minute comparison with the humeri of the aurochs or bison, but from its size it may be safely referred to the *Urus priscus*, as that appellation is employed by us. It is thicker and shorter than the humerus of the adult aurochs in the British Museum. None of the fossil radii in the collection at Haslar Hospital are small enough to fit the condyles of this humerus.

	NO. 81.	NO. 99.
Length from tip of greater tubercle to the distal corner of the outer condyle	13·5 in.	— in.
Length from convexity of the head to middle trochlea of the condyles .	11·9	—
Greatest transverse diameter of the proximal end	5·4	—
Ditto ditto of the distal end	3·7	3·4
Circumference of the shaft where it is smallest	7·0	4·8

No. 80. Lower part of the *left humerus*, slightly larger than preceding, and much decayed.

Depth or length of anconal fossa, including middle trochlea of the condyles   .   .   . 3·0 inches.

No. 99. A *left humerus* of a younger individual, wanting the head. This bone, though as slender in the shaft as the humerus of a musk-cow, has a much larger elbow-joint and a deeper and longer anconal fossa. The shaft also is less compressed, and the longitudinal range on its thenal side is much less evident.

A *tibia* collected by Captain Beechey, and figured by Dr. Buckland (pl. iii. f. 4), measures by the scale on which the figures are drawn 18·2 inches in extreme length, or about three-quarters of an inch more than the tibia of the aurochs whose dimensions are recorded by Cuvier in the ' Ossemens Fossiles' (vol. iv. p. 139).

Another *tibia*, half an inch shorter, but somewhat stouter, forms part of Captain Kellett's first collection, now in the British Museum. There is also a tibia numbered 248 in the same collection, of a larger size. All these were only cursorily looked at, time not permitting me to make a careful comparison between them and the corresponding bones of the aurochs or American bison.

BISON CRASSICORNIS. *Heavy-horned fossil Bison.* (Plate IX.)

No. 1 A and 4 r. The fragment of a *skull* brought home by Captain Beechey, and figured by Dr. Buckland in pl. iii. f. 1 (op. cit.), is referred by him to *Bos urus*, by which is meant the aurochs, or *Bison priscus* of more recent palæontologists. On comparing this with the skulls collected by Captain Kellett in the same locality, and which have been provisionally ranged above under the name of *Bison priscus*, it appeared, independently of its size, to possess characters by which it might be distinguished from them, and though these characters are not very striking, the distinctions between the skulls of the aurochs, mustush, and fossil *Bison priscus* are scarcely more pronounced ; a proper appellation has therefore been given to the fragment, by which it may be known until further collections shall furnish the means of deciding whether it be a relic of a distinct race of ancient bisons, or that the characters it presents depend on sex, age, or other individual peculiarity. A front view on a small scale is given by Dr. Buckland, as above mentioned, and one of half the natural size, also in front, but at a somewhat different line of sight (to correspond with those in Plate VII.), is represented in Plate IX. fig. 1. The profile, also of half-size, is given in fig. 4, and an occipital view of full size in fig. 2, with the basilar aspect of the condyles and basi-occipital in fig. 3. The occipital view, when contrasted with the similar ones of the aurochs and mustush in Plate VI., exhibits a marked difference in outline from either, and a greater breadth at the reverted edges of the squamosals, or the mastoid angles, as they are termed by Cuvier. Dr. Buckland's figure, when compared with one of Cuvier's (Oss. Foss. pl. xii. fig. 2), which is drawn on a like small scale, seems to present a striking resemblance. The original of Cuvier's figure was found by Mr. Peale in Big-bone Lick, on the Ohio, and from the plaster cast which was sent to Paris Cuvier pronounced it to be similar to the fossil bison skulls of Europe, but to be one of the very largest size (" *l'un des plus énormes*"). The height of its occiput from the inferior edge of the occipital foramen to the occipital crest is 7 inches; the vertical diameter of its horn-core also 7 inches; and the circumference of the latter at the base 21 inches[*].

The dimensions of Captain Beechey's specimen are :—

Distance from the basal ring of one horn-core to that of the other . . . . 11·6 inches.
Height of occiput from the lower edge of the occipital foramen to the centre of the
    occipital ridge . . . . 6·0
Distance between the most lateral edge of one squamosal to that of the other . . 11·0
Circumference of the more perfect horn-core at its base . . . . . . 12·6
From the base of the horn-core to the edge of the orbital plate, where the distance is
    least . . . . 3·8
Breadth of the forehead at the incurvatures between the orbital plates and bases of the
    horn-cores . . . . 10·9
Distance between the horn-cores to the nearest part of reverted edge of the squamosal,
    or width of the temporal fossa at its narrow point . . . . 1·2
Transverse distance between the outsides of the condyles . . . . . 5·7
Length from occipital arch to broken edge of frontal, about the junction of the nasals . 9·7

In the transverse arch of the forehead, and in the curve at which the orbital plates join the middle of the frontal, there is a likeness to the male mustush skull, which it also resembles in having a protuberance on each side of the sagittal suture, on a line with the posterior roots of the horn-cores. The orbital plates are like those of the aurochs, but the horn-cores have more the direction of those of the mustush. They are more depressed at the base or flattened on the concave side than those referred above to *Bison priscus?* and they are directed horizontally with a slight basilar inclina-

---

[*] Cuvier adds, that should a horn-core found by John Mayer in Bohemia prove to be of this species, which it resembles in its curves and in the forms of the portions of the skull adhering to it, it is of still greater dimensions than the American one, having a diameter of 7·9 inches at the base. The *Bos* (*Bison*) *bombifrons* of Harlan, described and figured by Dr. Caspar Wistar in the Transactions of the American Philosophical Society of Philadelphia (vol. i. new series, p. 375, pl. xi. f. 10, 11), is a quite different species, also disinterred at Bigbone Lick. Its frontal resembles that of a sheep in its transverse elevation between the horns.

tion, and more iniad, much in the way that the horns of the mustush would curve were that animal horned on a much larger scale.   Their backward position is such that a spot on their posterior edge, two inches from their base, is even with the sides of the occipital arch when the skull is seen in profile.   Though the cores are much wasted by decay, they are still considerably larger than those of an adult mustush or aurochs bull.

The angle of the temporal fossa formed by the base of the horn-core and the reverted occipital edge of the squamosal is narrow, not only comparatively but positively, being absolutely narrower than those of the adult aurochs or mustush in the British Museum, though the skull is larger.   The occipital plate resembles that of the mustush rather than that of the aurochs in its general surface and in the form of the rough triangle representing the spine, which, though it is not so acutely triangular as in the mustush, is dissimilar in shape to the corresponding spot in the aurochs.

In the outline of the occiput this skull is distinguished from that of the aurochs as well as of the mustush by its squamosals being less arched, as may be perceived by comparing Plate IX. fig. 2 with Plate VI. fig. 1 and 3.   In profile the occipital condyles fall more sacrad of the summit of the arch than in the aurochs, and still more than in the mustush, though not more than in *Bison priscus?* of the ice-cliffs.   Compare the fossils, Plate IX. fig. 4 and Plate X. fig. 1 (*priscus?*) with the living species represented in Plate VII. fig. 3 and 5, all in profile, and of half the linear size.

The condyles, Plate IX. fig. 3 (natural size), have a lateral prolongation not so well developed, nor assuming such a distinct trochlear form as in the mustush bull, and being rather intermediate between the form they present in that animal and that which they exhibit in the aurochs.   The absence or defect of this lateral trochlea indicates a corresponding difference in the brim of the antarticular surface of the atlas.

The basi-occipital is much wider on its basilar surface than that of *Bison priscus?*, No. 24,576 (Plate VI. fig. 5), its width being greater than its length, while in *B. priscus?* the breadth and length are nearly equal.   In the living aurochs also this bone has a broad, flat, basilar surface, different from that of *B. priscus?*, but not so broad and short as in *B. crassicornis*.   In the mustush the basi-occipital is comparatively narrow.

No vestige of a hypapophysial ridge can be detected on the basi-sphenoid, which is somewhat worn; neither is this visible in that bone of the mustush, though it is distinct in the aurochs, and still more so in the domestic ox.

From the sagittal suture traversing the forehead to between the horns we infer that the skull of *B. crassicornis* here described was not that of an aged individual.

No. 91.   This number indicates the *large horn-core*, of which a side view on the facial aspect is given in Plate XIII. fig. 1, and a view of the coronal aspect in fig. 2, both of the natural size.

The circumference at the widest part near the base is .   .   .   .   . 15·2 inches.
            „   immediately adjoining the broken edge towards the tip .   . . 10·8
            „   of the neck formed by the frontal .   .   .   . 13·5
The vertical diameter at the thickest part is   .   .   .   .   . 4·2
And the antero-posterior one   .   .   .   .   .   . 5·2

The fragment weighs 7⅓ lbs. avoirdupois, and is much too massive to resemble the round, light, tubular cores of our large domestic oxen; and the cores of the *Bos primigenius* are, as far as one may judge by the figures that have been published, compressed at the base in a different direction.   In the fragment of the frontal from whence the core springs there appears to be evidence of its having been broken from a skull of the bison type, but its size and form point it out as belonging to a different species from the one with which it is contrasted in the plate, and which has been referred to *Bison priscus* (p. 35); it has therefore been considered a horn-core of an older and probably a male individual of the race that produced the skull marked No. 1 A, and to which, from the thickness of its horns, I have given the distinctive epithet of *crassicornis*.

Cuvier, whose determinations are entitled to every consideration, has referred several horn-cores of great size to the European fossil bison, and among others one described by Faujas, and figured in pl. xii. fig. 4 of the ' Ossemens Fossiles;' also one of doubtful species, described by John Mayer; one of them having a circumference at its base of 13·4 inches, and the other of 23½ inches. Peale's specimen, dug up at Bigbone Lick, and referred to above, has a horn-core which measures more than 21 inches round; and the British Museum contains one numbered 21,304, which was dug up at Grays on the Thames, measures 19 inches in circumference, and is very similar to No. 91, which we are now considering. There is therefore abundant authority for ranging under the head of *Bison priscus* horn-cores so different in size and dissimilar in appearance as those represented in Plate XIII.; but finding among the bovine vertebræ dug up in the ice-cliffs at least two forms allied to the bisons, besides two clearly appertaining to the musk-ox type, I have been led to seek for peculiarities in the crania and horns, by which an equal number of species might be characterized. And moreover pl. xi. of Meyer's paper shows that the cores of *Bison priscus* of Europe may acquire a very great length without any extraordinary thickness. Core No. 91 is very flat transversely, or depressed on its concave or coronal surface, in comparison with the fossil bison one represented on the same plate (fig. 3), and its inial side has a convex outline, while the antinial side is straight. The basilar aspect, or that of the longitudinal convex curve, is convex also transversely, forming a segment of a circle. The part of the frontal that projects to form the base of the core is narrower than the adjoining part of the proper core, and produces as it were a neck, which appears concave in its profile when we look directly at the flat side of the core, as in fig. 2, and concave also but more flatly on the basilar surface. In being crossed by a smooth groove and in the general appearance of the surface of its basilar side No. 91 much resembles the same part of the *Bison priscus?* of Eschscholtz Bay.

The core has been exposed to much friction, and when entire must have measured more in circumference than at present. It still retains the remains of longitudinal grooves, more numerous on the basilar aspect, but also existing on the flat side, particularly towards the narrow end. A thick net-work of strong walls forms closed cells in the interior, and does not admit the light to traverse it. Its cellular structure is similar to that of the *Bison priscus?* cores.

No. 90. This number is attached to an *atlas* in Haslar Museum, which was found on Captain Kellett's second visit to the ice-cliffs. It has lost much of its animal matter, is decayed on the surface, and the anterior parts of its lateral processes are broken off. It is figured of the natural size in Plate XII. Its antarticular cavity is slightly larger than that of an atlas which would exactly fit the fragment of the skull of *Bison priscus?* (No. 140) mentioned in page 36, and too small to receive the condyles of the skull I have provisionally named *Bison crassicornis*.

	NO. 90.	MUSTUSH.	AUROCHS.	DOM. OX.
Transverse axis of the brim of the antarticular cup .	5·4 in.	4·65 in.	4·5 in.	4·65 in.
Sterno-dorsal axis of ditto	2·6	2·0	—	2·2
Length of centrum, mesial line of sternal surface .	2·0	1·95	1·75	1·8
Length of vertebra, from edge of prozygapophysis to the sacral extremity of the pleurapophysial process	5·5	4·8	4·7	4·6

G 2

	NO. 90.	MUSTUSH.	AUROCHS.	DOM. OX.
Lateral axis of post-articulating surface . . .	5·0 in.	4·2 in.	4·1 in.	4·4 in.
Distance between the peripheral summits of neural and hypapophysial spines	4·3	3·9	4·3	3·8
Transverse distance from the most lateral part of one pleurapophysis to that of the other . . .	7·9	7·5	7·6	8·0
Transverse diameter of proximal end of vertebra, including the thickness of the brim	6·6	—	5·5	5·3

This vertebra, independent of its greater size, is more stoutly framed than the corresponding one of either the mustush or aurochs. A reference to Plate X. will make known its general difference of form from another fossil atlas, from the same locality, but of a slighter construction; and Plate VIII., which contains views of the atlas of the mustush and of the aurochs reduced to half the natural size, will show that they are differently formed in many particulars from either fossil. Because this atlas seems to be the best adapted by its great strength for the support of a head, rendered more than usually heavy by a great mass of horn core and case, I have ranged it with the skull of *Bison crassicornis*, and the brim of the articulating cup also seems to correspond with the form of the condyles of that skull. This arrangement, however, is to be considered as open to future investigations.

Atlas No. 90 differs from that of the mustush in the brim of its antarticular cup, wherein it has a closer resemblance to the atlas of the aurochs. This brim is scarcely deficient, or at least but very slightly notched at the confluence of the articular surfaces of the centrum and prozygapophysis, resembling in that part the atlas of the aurochs; but the border of the very superficial notch that does exist is so broad as to afford sufficient room for an articulating facet, on which the lateral prolongation of the condyle may play; whereas in the living species the edge is acute, and the condyle ending abruptly is received entirely within the articular cup. In the mustush the notch is very conspicuous, and its edge is broad and flat, furnishing a smooth articular surface, on which the accessory lateral trochlea of the condyle moves. The surface of the fossil No. 90 is too much worn to show any remains of the glazing of such a facet, supposing it to have existed. The notch over the neural canal between the prozygapophyses is shallower and wider than that of the aurochs, which again is less deep than in the mustush, and differs strikingly from the acute notch in the fossil atlas of our *Bison priscus?* figured in Plate X.

The wide notch, or confluent pair of notches, which exists on the sternal side of the antarticulating cup, and admits of the shoulders of the basi-sphenoid abutting against the proximal edge of the centrum, is formed like that of the aurochs, but is more rounded off on the edges, as indeed it requires to be, from the much greater thickness of the bone. In the mustush this articulating surface is more sharply defined, and flatter in one sense, but the whole notch is proportionally deeper. The narrow process which stands up in the atlas of the last-named species, between the notch in question and the lateral accessorial facet, by its form at once distinguishes it from the corresponding vertebra of the aurochs or from that of the fossil No. 90.

The centrum, as distinct from the lateral processes, presents a broad sternal surface, which is abruptly defined on each side, by deep depressions in these processes, situated beneath and including the arterial foramina. In the domestic ox the depressions are wide and shallow, the substance of the bone being much less, and they run out gradually on the sides of the centrum, rendering that part narrower.

The hypapophysial tubercle is much stouter than in the male aurochs, but is not more prominent, and not so near the post-articular surface as in the mustush or domestic ox. The neural tubercle or spine is well shown in Plate XII. fig. 3, and is much stronger than that of the aurochs.

The lateral processes are mutilated only at their distal extremities, which are thick and stout, and droop considerably sacrad. They embrace a greater portion of the margin of the post-articulating surface of the vertebræ than in the domestic ox, wherein from their greater thinness a projecting ledge of the articulating surface is formed on the neural side. In the fossil this ledge is nearly obscured by the thickness of the bone which supports it. The sacral aspect of the lateral processes is therefore, as a consequence of this thickness, broader and flatter than in either the aurochs, mustush, or domestic ox.

In the interior of the neural canal, the centrum is crossed transversely by a smooth groove big enough to hold the little finger, as shown in Plate XII. fig. 4, which ends by a deep pit in each neurapophysis, immediately sacrad of the arterial foramina, and just above the edge of the odontoid process of the second vertebra. These depressions exist in the atlas of the domestic ox, but the connecting groove is wanting, there being merely some inequalities for the attachment of the *ligamentum vitæ;* neither in the aurochs is there a groove at all approaching in depth and distinctness to that of this fossil. There is, however, a shallow groove in the mustush.

On comparing this fossil with an atlas found at Grays on the banks of the Thames, its distal articular surface was found to be less oblique, and the form of the lateral processes different. The extreme width of the Grays atlas is 11·3 inches.

Three of the following cervical vertebræ and the first dorsal (Nos. 115, 116, 120, 121) belonged unquestionably to one individual, and are as well preserved as they could have been had the animal died within a month or two. They are of a light brown colour, their surfaces being glossy as if slightly varnished, and without the least appearance of friction. Little of their animal substance has perished, the bones being merely somewhat more fragile than when quite recent, and the articulating surfaces still show portions of the fibro-cartilaginous matter which lined them when alive. In respect to the length of the centra, and consequently of the spinal column generally, they are about equal to the corresponding bones of our largest domestic oxen, but all their parts are much more robustly made and their strength vastly greater. It is probable that the fossil animal had a shorter and stouter neck, for the support of a heavy head and horns. In most of their dimensions these cervicals considerably exceed those of the full-grown American mustush or Lithuanian aurochs. Unfortunately we have no certain data for connecting them with any of the skulls obtained from the cliffs; as they are however the largest in the collection they may be provisionally referred to the *B. crassicornis*, indicated by the skull brought home by Captain Beechey, figured in the Appendix to his voyage (pl. iii.) and in Plate X.

In the domestic ox the cervicals have the distal end of the hypapophysial ridge projecting sacrad, giving a peaked corset form to the centra and an obliquity to the distal articulating cup, which exists in a very much less degree in the fossil vertebræ from Eschscholtz Bay, and in those of the mustush or aurochs, and not at all in those of the musk-ox. This obliquity is not absent in the sixth and seventh cervicals of the ox, which have no hypapophysial ridge; and an approach to it in that species is seen even in the atlas by the hypapophysial tubercle projecting over the edge of the distal articulating cup.

No. 115. *Third cervical.*—The neural spine, portions of the transverse processes, and the distal edges of the zygapophyses are broken off, in part since the bones were disinterred, but in all other respects the condition of the specimen is perfect, except that the left parapophysis has been fractured during life and re-united by means of much callus with the diapophysis.

	FOSSIL.	DOM. OX.
Length of centrum on its sternal aspect, from its antarticulating crown to the brim of post-articulation	4·3 in.	4·2 in.
Length of centrum on the same aspect, excluding the antarticulating ball . .	2·6	2·8
Length of centrum on its dorsal side, excluding the antarticulation . . .	3·1	2·7
Ditto from proximal edge of prozygapophysis to distal edge of zygapophysis .	4·4	3·8
Sterno-dorsal diameter, from tuberosity of hypapophysis to crown of neural arch, between zygapophyses	3·8	3·6
Sterno-dorsal diameter, from sternal side of antarticulating ball to crown of neural arch, between the prozygapophyses	3·4	2·9
Lateral diameter, measured between the prozygapophyses and lateral processes, outside of the arterial foramina	2·6	2·4
Lateral diameter, from the lateral edge or shoulder of one prozygapophysis to that of the other	4·1	3·2
Lateral diameter, from lateral edge of one zygapophysis to that of the other .	4·7	3·2

When compared side by side with the corresponding bone of the domestic ox, it immediately strikes the eye as being much stronger built; and as the preceding table of dimensions will show, its greater dimensions are chiefly owing to the robustness of its several parts, and not to the absolute length of the centrum. If the neck were of the same proportional length in the two species, the fossil bone might have been assigned to a stout bull, whose length did not exceed one of our large Lincolnshire oxen; but as the collection contains also three lumbar vertebræ, most probably part of the same skeleton to which this cervical belongs, and these are every way bigger than those of any of our oxen or of any existing aurochs or mustush, we seem justified in attributing greater bulk to the whole frame of the fossil animal, and this would be in accordance with the dimensions of the skull, which we feel inclined, as we have said above, to associate with these cervicals and lumbars.

The antarticulating ball differs in form in the two species, that of the fossil being proportionally larger, wider on the neural side, with a larger sterno-dorsal axis, less flattened laterally, and more elliptical in outline on the sternal aspect. The hypapophysial ridge is stouter in the fossil, not so acute, and its termination sacrad is wider and blunter, but equally prominent. In the domestic ox, up to the age of four years, the autogenous post-articulating epiphysis of this cervical has not coalesced with the centrum, and a peaked process of it, partly cartilaginous, may then be seen interposing between the sides of the hypapophysial tuberosity: in the musk-bull cervical, which has no hypapophysis, this peaked process is wanting, and as far as we are able to make out, it seems to be absent or nearly so in the fossil cervical now specially alluded to, but coalescence and ossification have proceeded too far to allow the matter to be absolutely determined. The post-articulating cavity of the centrum is shallower in the fossil, rather wider in both its dimensions, and its brim is considerably less oblique, from the hypapophysis terminating before reaching it, instead of jutting out a quarter of an inch beyond it, sacrad, as in the domestic animal. The neural canal is larger in the fossil; but the neural spine, as far as can be ascertained from the fragment that remains, was little, if anything, larger than in the domestic animal.

The prozygapophyses are much more massive than in the domestic ox, their articulating surfaces are very flatly concave, and show but faint traces of the two convex surfaces, separated by a longitudinal hollow, which is visible enough in the corresponding process of the domestic animal. The backs of these apophyses are very convex, being segments of spheres, so that the most laterally prominent part is not the edge of the articulation as in the ox, but the bulging shoulder that supports it. This shoulder projects beyond the atlantal edge of the articulation, forming a crest which can be traced along the edge of the neural arch to the base of the spine. There is a minute indication of this ridge in the cervical of the ox, and it is prominent and greatly developed in the musk-bull, being probably what Professor Owen names the metapophysis.

Zygapophyses equally strong and convex have a greater comparative lateral expansion in relation to those of the prozygapophyses than in the ox, so that the bony plate which forms the connecting link between the oblique processes joins the back of the zygapophysis at some distance from its border instead of close to it, as in the ox. There are other rough lines and eminences on the back of the zygapophyses, which indicate the play of more powerful muscles than in the domestic animal, in which this part is nearly smooth. The large zygapophyses are separated by a deeper and more acute notch than in the ox, the difference in the two species with reference to a line drawn across the neural canal being nearly an inch and a half.

With respect to the lateral process formed of the coalesced parapophysis, pleurapophysis, and diapophysis, that of the fossil being in part broken off, its precise form cannot be described. It has however a general resemblance to that of the domestic ox, the parapophysis projecting atlantad with a curve, as in the ox; but the edge of this process, instead of running almost straight sacrad on the sternal side of the diapophyses to its distal end, terminates suddenly about an inch from it, while the pleurapophysial end of the diapophysis takes a lateral course.

No. 116. This is known to be the *fourth cervical* of the same individual by its similarity of condition and by its fitting very exactly in its antarticulations with the third one (No. 115); and its differences from the corresponding parts of the same bone of the domestic ox are of a similar kind with those particularized in the description of No. 115. The metapophysis is a tubercle not higher but thicker and even more distinct from the articulating surface of the prozygapophysis than in the third cervical. On the edge of the lateral process there is seated a rough tubercle, which occupies nearly the whole lateral space between the prozygapophysis and the zygapophysis. In the domestic ox this tubercle is much less prominent, lies nearer the edge of the zygapophysis, and does not approach so near to the prozygapophysis. The notch between the zygapophyses in No. 116 is even deeper than in the third fossil cervical, and is of the same acute form, and the angle which the articulating surface of the prozygapophysis makes with the mesial plane of the neural spine is less acute than in the domestic ox.

The parapophysis differs from that of the ox in form, being not only more substantial, but having a much broader surface on its mesial aspect, and a different profile, not cut away in a straight line on its sternal side, but rounded. Its exact form cannot be described, however, as its edge has been broken. The diapophysis is also broken at the end, but its base is much stronger than in the ox. A pit separating the distal edges of the two processes is similar in the two species, but the edge of the parapophysis is prolonged along the side of the centrum, nearer to the distal brim of the post-articulation than in the ox. Other differences may be gathered from the annexed table of dimensions.

The rough tubercle on the base of the zygapophysis and the edge before it is scarcely so much developed in the aurochs as in the ox, and between that process and the prozygapophysis there is a smooth and pretty deep notch, three-quarters of an inch wide. The centrum of the fourth cervical

of the aurochs presents some differences also in its hypapophysis, and the notch between the zygapophyses is less deep.

	FOSSIL.	DOM. OX.
Length of centrum on its sternal aspect from its antarticulating crown to the brim of post-articulation	4·2 in.	4·2 in.
Length of centrum on the same aspect, excluding the antarticulating ball .	2·4	2·8
Length of centrum on its neural side, excluding antarticulation . . .	2·8	2·5
Length from proximal edge of prozygapophysis to distal edge of zygapophysis .	4·2	3·9
Sterno-dorsal diameter measured from the tuberosity of the hypapophysis to the crown of the neural arch between the zygapophyses	3·8	3·6
Sterno-dorsal diameter from the sternal side of the antarticulating ball to the crown of the neural arch between the prozygapophyses	3·2	2·7
Lateral diameter measured from the outer walls of the arterial foramina between the prozygapophyses and the lateral processes	3·2	2·6
Lateral diameter from the lateral edge or shoulder of one prozygapophysis to that of the other	4·9	3·4
Lateral diameter from the lateral edge of one zygapophysis to that of the other .	4·6	3·5

No. 117 is a *fifth cervical*, more worn and decayed than Nos. 115 and 116, having lost all the
gloss and much of the original surface. It evidently belongs to the same species, but to a different
and slightly larger individual.

As contrasted with the corresponding cervical of the domestic ox, the hypapophysial keel appears blunter and at the same time more boldly relieved from the rest of the centrum by deeper depressions between it and the parapophyses; the keel has less rake, or projects less sternad towards its
distal end, and as a consequence of this the post-articulation is less oblique.

The parapophysis is differently shaped from that of the ox, being flatter, with a greater atlanto-
sacral diameter, and a more straight sacral or sterno-sacral edge, which wants the projecting angle of
the corresponding process in the ox. There is a deeper depression between the parapophysis and diapophysis, and the latter is differently formed and differently united to the parapophysis than in the ox.
These peculiarities of the parapophysis and diapophysis also exist in Nos. 115 and 116, but this bone
being more complete on one side, their forms are more perfectly shown in it. No. 116 does not fit
this cervical, there being a difference in the inclinations of the oblique articulations. In the acuteness
and depth of the notch between the zygapophyses this fifth cervical agrees with the two preceding
ones (Nos. 115, 116). The prozygapophyses are strengthened by the addition of much convex bony
matter on their outsides, and a distinct small angular process or metapophysis rises atlantad of the
articulating surface.

	NO. 117.	DOM. OX.
Length of centrum on its sternal aspect, including the antarticulating crown .	4·3 in.	4·1 in.
Ditto ditto ditto excluding the crown . . .	2·9	2·6
Ditto ditto on its neural side, excluding the crown	2·7	2·5
Length from proximal edge of prozygapophysis to distal edge of zygapophysis .	4·1	3·5
Sterno-dorsal diameter from sternal side of antarticulation to crown of neural arch between the prozygapophyses	3·1	2·7
Lateral diameter measured in the interval between the prozygapophyses and diapophyses outside of the arterial foramina	3·2	2·9
Lateral diameter between the outsides of the prozygapophyses	4·8	3·7
Lateral diameter between outer edges or tubercular shoulders of zygapophyses .	4·4	3·8

There is a peculiarity in the connection of the diapophysis with the brim of the articulating cup of this cervical. In the domestic ox the arterial foramen opens exactly between the two, shortening the diapophysis, while in the fossil the foramen is more dorsad and the base of the process curves down to the brim. In the distal arterial foramen of the fourth cervical opening dorsad of the diapophysis, and at some distance from the post-articulation, the fossil animal and the domestic ox agree; and in the third cervical the diapophyses and parapophyses, having a common distal base, in both species, sternad of the arterial foramina, do not furnish a distinguishing character, though in the fossil the foramen opens on the edge of the articulation, while in the domestic ox its aperture is more atlantad.

The fifth and sixth cervicals of the individual to which the third and fourth described above belong do not exist in the collection*.

No. 120. The *seventh cervical* was obtained, however, complete in every respect except the ends of the neural spine and diapophyses, and in the same excellent state of preservation with the third and fourth, with the single exception of its being slightly coated in some parts with phosphate of iron. The distal surface of this fossil is figured in its proper dimensions in Plate XI. fig. 6, and the remarkable strength of the neural spine distinguishes it at once from the same bone in the domestic ox, aurochs, or mustush.

The centrum differs from that of the corresponding bone of the domestic ox, in its sternal surface being less gradually rounded off laterally and more decidedly bounded on each side by an obtuse rising, that ends sacral in a stout protuberance, contiguous to the articular saucer for the reception of the head of the first rib. This saucer, the tubercle in question, and a smaller one connected with it on its mesial side but situated more sacral, all belong to the distal autogenous articulating process, which does not coalesce with the centrum till after the fourth year. The coalescence is perfect in the fossil, and all the tubercles are more prominent in it than in the domestic ox. They are shown, in profile, in fig. 6. The lateral tubercles of the aurochs are similar to those of the fossil, but they scarcely appear in the mustush.

The antarticulating head of the centrum has a proportionally longer sterno-dorsal diameter than in the ox, and the post-articulating cup a proportionally shorter one.

The zygapophyses of the fossil stand more dorsad from the centrum in comparison with the prozygapophyses than in the domestic ox, so that when the seventh cervical of the latter is viewed at the same angle at which the fossil is shown in fig. 6, the edges of the prozygapophyses hide only part of the zygapophyses.

In the domestic ox the neural spine of this vertebra is two-sided, with the sacral edge a little more obtuse than the very thin dorsal one, and the lateral diameter of the bone is increased in a moderate degree near its sacral edge. In the fossil, however, the base of the spine is wide and concave on its sacral aspect, buttressing the zygapophyses for their whole breadth; it then swells up, forming a rough prominent tubercle, beyond which, had the spine been entire, we should most probably have found a thin edge. In the domestic ox the height of the spine scarcely exceeds the distance from its base to the sternal side of the centrum; but it has evidently been much longer in the fossil animal.

The metapophysial crest of the prozygapophysis, which is very evident in the third and fourth

* In the British Museum there is a sixth cervical, numbered 35, in good condition, said to have been part of Captain Beechey's collection, and it is of even a larger size than the preceding, and equal to one of the largest cervicals dug up at Grays on the Thames, but as it is not mentioned in Mr. Collie's list its origin must be considered doubtful.

H

fossil cervicals, becomes nearly obsolete in the seventh. It is a small cone in the seventh of the musk-bull.

A small rough line seen in the fossil crossing the junction of the diapophysis with the prozygapophysis does not appear in the vertebra of the ox; and the rough spot between the zygapophysis and prozygapophysis is much broader than in the domestic ox, and not so tubercular as in the third and fourth fossil cervicals.

On the right side of the fossil a bony bridge has been formed from the lateral process to the centrum, converting the notch for the exit of the nerve into a foramen. This is represented in fig. 6 as occurring on the left side, which arises from the drawings on the stone not having been reversed, so as to give a right impression.

	NO. 120.	NO. 122.	OX.
Length of centrum, from the crown of its antarticulating ball to the brim of its post-articulation on the sternal side	3·4 in.	3·2 in.	3·3 in.
Length of centrum, same side, excluding the antarticulation . .	2·0	1·9	1·7
Length of centrum, neural side, excluding its antarticulation . .	2·3	2·1	2·0
Length from proximal edge of prozygapophysis to distal edge of zygapophysis	4·4	4·2	3·5
Sterno-dorsal diameter, from the brim of post-articulation to the crown of the neural arch, between the prozygapophyses . . .	3·6	3·4	3·2
Lateral diameter, measured in the interval between the prozygapophyses and diapophyses	4·1	3·8	3·3
Lateral diameter, from the edge of one prozygapophysis to that of the other	4·7	4·5	4·2
Lateral diameter, from the lateral edge of one zygapophysis to that of the other	3·6	3·3	3·4
Lateral diameter of the centrum at the articulations of the heads of the ribs	3·5	3·4	3·2
Vertical diameter of neural canal, where it is least	1·1	1·1	1·0
Transverse diameter of the neural canal	1·3	1·2	1·4

No. 122 is a *seventh cervical*, smaller, but similar in all points of form to No. 120. It is much more weathered, and has lost more of its animal matter.

No. 119. *Seventh cervical.* This fossil is much more decayed than No. 120 and the others belonging to that skeleton, having lost very much of its animal matter, and its surface having acquired a fibrous aspect. It differs from No. 120 in the somewhat greater inclination outwards of its prozygapophyses, and in having a more slender neural spine somewhat differently impressed at the base between the zygapophyses; but these differences rather incline us to consider it as having belonged to a female of the same race, than to a different species. It is not the relic of an aged animal, as the articulating epiphysial cup of the centrum shows signs of separation.

Length of centrum, sternal side, including articulating ball	3·5 inches.
Ditto ditto excluding articulating ball	2·0
Ditto ditto neural side, ditto	2·2
Lateral diameter of centrum, between the outer edges of the cups for the heads of the ribs	3·3

Sterno-dorsal diameter, from sternal side of centrum to crown of neural arch, distal
    end . . . . 3·5 inches.
Transverse diameter, measured from the outsides of the prozygapophyses . . . 4·9
Transverse diameter, measured from the outer sides of the zygapophyses . . . 3·8
Transverse diameter, measured at the intervals between the prozygapophyses and lateral
    processes . . . . 4·2
Vertical diameter of neural canal where narrowest . . . . . 1·3
Transverse ditto   ditto . . . . . . 1·3
Atlanto-sacral diameter of neural spine at its base . . . 2·2

No. 121. This *first dorsal* evidently belongs to the same individual skeleton with the seventh (No. 120) and the third and fourth cervicals (Nos. 115 and 116) above described. The colour and condition of their surfaces, which are peculiar, are precisely the same, and the prozygapophyses of No. 121 fit exactly to the zygapophyses of No. 120. If tried with No. 119, which differs little in size from No. 120, the joints do not coincide, owing to the greater obliquity of the zygapophyses of No. 119. The fossil is defective in the end of the spine, the transverse processes, and the sternal side of the centrum, which have been broken off; elsewhere it shows no signs of decay, and the remains of the synovial membrane still line part of the articulating surfaces. The centrum is larger than that of a full-sized Lincolnshire ox, the neural spine very much stouter, and exceeding in thickness that of an adult aurochs, and still more that of the mustush.

When compared with the first dorsal of the domestic ox, the position of the prozygapophyses furnishes a conspicuous differential character. In the fossil the articulating surfaces of these processes extend mesiad over the neural arch towards the root of the spine, so that the distance between them scarcely exceeds half an inch, while in the ox they are more than an inch and a quarter apart, and scarcely encroach on the neural arch at all. In the fossil they are quite flat, while in the ox they are convex, both transversely and longitudinally.

The form of the centrum cannot be made out; but judging from the small portion that remains behind the articulation for the head of the first rib, it would appear that the centrum was less compressed or hollowed laterally than in the ox.

The neural spine is more than 1½ inch wide transversely at its base, immediately adjoining the zygapophyses, and 2½ inches in atlanto-sacral diameter at the same place; and the articulating surfaces of the zygapophyses are in close contact with each other, while in the ox they are separated from one another by a groove.

	FOSSIL.	LARGE OX.
Length of centrum, from the crown of its articulating ball to the brim of its post-articulation, on the sternal side	— in.	3·4 in.
Length of centrum, same side, excluding articulation	—	2·2
Length of centrum, dorsal side, excluding antarticulation	2·5	2·0
Length from the proximal edge of the prozygapophysis to the distal edge of the zygapophysis	4·1	3·8
Lateral diameter from lateral edge of one prozygapophysis to that of the other	3·5	3·7
Lateral diameter from lateral edge of one zygapophysis to that of the other where widest	1·9	1·4

	FOSSIL.	LARGE OX.
Lateral diameter of centrum, at the articulation of the head of the first rib, on its dorsal border	2·9 in.	2·7 in.
Vertical diameter of neural canal	1·4	0·9
Transverse diameter of neural canal	1·4	1·2

There are no other dorsals that can be assigned to the same skeleton to which the cervicals and first dorsal above described belonged, but two vertebræ much mutilated may be the remains of a younger individual of the same species, if we ought to judge from their relative size, which is greater than that of the corresponding vertebræ of either the aurochs or mustush.   One of these,

No. 129, is a *sixth or seventh dorsal,* with the walls of the neural canal, the base of the neural spine, and the dorsal half of the centrum remaining, but the articulating epiphyses of the centrum mostly gone.   It fits No. 130 so exactly, and the colour and fossil condition of both are so much the same, that we cannot err in ascribing these two to the same individual, but, from the manner in which the epiphyses have separated, to a younger animal than that to which Nos. 115, 116, etc., belonged. In general aspect both No. 129 and No. 130 have a nearer resemblance to the dorsals of the mustush than to those of the aurochs, and especially in their articulating surfaces; the metapophysis does not appear to have been more conspicuous than in an aged domestic cow, but none of these parts are quite perfect in the fossil.   The sternal side of the centrum of No. 130 is narrow, but not so acute as in the domestic cow; and the mesial ridge of the centrum, within the neural canal, is neither prominent nor acute.

	NO. 129.	NO. 130.
Length of centrum, from the crown of the articulating ball to the brim of its post-articulating cup, on the sternal aspect	— in.	3·0 in.
Length of centrum, excluding the articulations	—	2·5
Sterno-dorsal diameter, from brim of post-articulation to the crown of the neural arch	—	3·2
Lateral diameter, measured from the outsides of the distal cups which receive the heads of the ribs	3·6	3·4
Lateral diameter between the outsides of the proximal cups for the heads of the ribs	—	2·7
Lateral diameter, from the outer edge of one prozygapophysis to that of the other	1·7	1·5
Lateral diameter between the outer shoulders of the diapophyses, about	4·8	—
Sterno-dorsal diameter of the spine at its base	2·4	—
Lateral diameter of ditto, about	1·6	1·6

Nos. 145, 146, 147, 150, and 199 are fragments of *ribs,* all of them much stouter than the ribs of the domestic ox, and not showing the thin and expanded blades which that species possesses; neither have they the character of the ribs of a musk-ox of large dimensions.   They are stronger than the ribs of the skeletons of the aurochs or mustush in the British Museum, and they are therefore ranged with the other supposed remains of *Bison crassicornis.*

No. 145 is the longest fragment, and measures about 18 inches in length, but has lost the two ends.   It appears to be part of the *fourth or fifth rib of the right side.*   Its breadth where widest, or not far from its sternal end, is 1·6 inch, and its greatest thickness, which is not far from the tubercle which articulates with the diapophysis, is one inch. At its widest part its thickness has diminished to 0·6 inch. At its greatest curvature, about two inches from the tubercle, and on its sacral edge, there is a rough swelling, which is much less distinct in the common ox, and does not appear at all in the musk-bull.

No. 147 is a shorter and stouter fragment, of apparently the *seventh rib of the left side.* The rough protuberance on its sacral edge, mentioned in the account of No. 145, is much larger and more prominent in this fragment. Its thickness, including that protuberance, is 1·7 inch, and an inch or two sternad of the protuberance it is still 1·5 inch. The breadth at the latter place is 1·2 inch.

The other fragments are portions of ribs of as great or even greater strength.

In the absence of an entire series of the dorsals or lumbars, we cannot determine whether the *Bison priscus?* or *crassicornis* resembled the mustush and aurochs in the numbers of their ribs.

Nos. 123, 124, 125, and 89 refer to three *posterior lumbars* and part of the *sacrum,* evidently members of one skeleton, and, if we may judge from size alone, of that individual skeleton to which the cervicals and first dorsal described above belonged; but as their fossil condition is somewhat different, this cannot be affirmed positively. Their pleurapophyses are broken, they have suffered some slight mutilations elsewhere, and they do not possess such a very fresh condition of surface as No. 115 and its fellows.

No. 123. A *lumbar,* being the third from the sacrum, and considerably bigger than the corresponding vertebra of the adult aurochs, and more like that of the mustush in general aspect, though on a larger scale. Plate XV. fig. 1 represents this vertebra as seen on its sternal aspect, and fig. 2 gives a side view of the natural size. The metapophysis seems to have been more depressed laterally, and less acute sacrad, than that of the mustush. The sigmoid curves of the zygapophyses are also deeper and more complete in the fossil. Means of drawing a comparison between these lumbars and the corresponding one, of a full-grown domestic bull of large size, have been wanting; but on contrasting them with the lumbars of an aged Alderney cow, much similarity in general form was seen to exist. The hypapophysial ridge of the fossil (No. 123) is scarcely so acute as in the cow, though it is more distinctly relieved from the rest of the centrum by longitudinal depressions along its base, and it has a more decided protuberance at its upper end. The metapophysial tubercle is more flattened on its dorsal and lateral aspects, and not so regularly convex as in the cow : it ends likewise sacrad in a different manner. A depression on each side of the neural spine, and immediately sacrad of the prozygapophysial articulation, distinguishes the fossil from the lumbar of the cow, and still more evidently from that of the musk-bull. A smaller fossil lumbar (No. 134), and probably a more anterior one, wants the depression in this place, but has a larger one nearer the zygapophysis. This second depression exists also, though less conspicuously, in No. 123.

The zygapophyses are proportionally larger in the fossil, and the dorsal edge of the joint curves over the edge of the receiving prozygapophysis. This greater development of the apophysial joints was pointed out by Bojanus as a character distinguishing the lumbars of the aurochs from those of the domestic ox. I have not been able to institute a correct comparison between the fossils and the corresponding bones of the adult aurochs in the British Museum, as the latter are parts of a mounted skeleton, and these joints cannot be distinctly seen. Though this excurvature of the mesial margin of the zygapophysis does not occur in the lumbars of the old Alderney cow, it exists in the last dorsal, and it may be seen in the second lumbar and last dorsal of the musk-bull, though not in any of the lumbars of the adult musk-cow. It would seem therefore that individual varieties, in respect to developments of the margins of the joints, occur independent of species. The sigmoid curve of the zygapophysial articulating surface exists distinctly in the musk-bull, domestic ox, aurochs, mustush, and in our fossil lumbars, but from the greater size of the processes and greater sharpness of the curves, they are more beautifully distinct in the latter than in any of the other species named.

In the fossil the zygapophyses project sacrad to a greater distance behind the base of the neural spine, and are separated from one another by a considerably narrower notch than in the ox.

The neural arch is a flat ellipse, and does not partake of the pointed form which it possesses in the musk-bull. In the latter the crown of the arch between the prozygapophyses is on a level with the upper curve of the sigma, and the mesial portion of the joint rises obliquely on the side of the arch in the direction of the neural spine; while in the fossil the mesial half of the joint lies almost horizontally over the comparatively flat roof of the arch, the whole of the joint being situated above the plane of the crown of that arch.

Within the neural canal the centrum presents a narrow ridge, which is prominent and very acute, between the two foramina.

Length of centrum on its sternal side, including convexity of antarticular surface .	. 3·3 inches.
Ditto ditto ditto excluding articulating surface .	. 3·0
Length from proximal edge of prozygapophysis to distal edge of zygapophysis	. 4·7
Length of centrum, dorsal side .	. 3·1
Lateral diameter of centrum at the notch sacrad of the roots of the diapophysis	. 2·3
Sterno-dorsal diameter to crown of neural arch, distal end .	. 3·3
Lateral diameter between the outsides of the metapophysial tubercles .	. 3·7
Ditto between outsides of zygapophyses .	. 2·4

The centrum of the corresponding lumbar of the mustush-cow, excluding the articulating surfaces, measures 1·85 inch in length.

No. 124. This *lumbar*, the second from the sacrum, differs from the corresponding one of the domestic ox in the same characters that were pointed out in the notice of No. 123. One of its prozygapophyses is almost perfect, so that the difference in the shape of its metapophysial tubercle may be fully made out. This tubercle consists principally of a great thickening of the process laterally on the outside of the joint, and from the most lateral part a small conical point is directed atlantad; on the sacral edge of the same surface there is a rough shoulder with a blunt edge, descending to the root of the diapophysis, but no ridge such as that visible in the cow. The notch between the zygapophyses is not narrower than that of the ox, in the same degree as in the preceding lumbar; and the neural spine is higher in proportion than in the ox; while the circumscribed pit on each side of the atlantal edge of the neural spine is rather deeper than in the preceding lumbar.

Length of centrum on its sternal side, including the convexity of the antarticular surface	3·4 inches.
Ditto ditto ditto excluding antarticulation .	. 2·9
Length of centrum on its neural side, excluding antarticulation .	. 3·2
Length from proximal edge of prozygapophysis to distal edge of zygapophysis	. 4·6
Lateral diameter of the centrum at the notch sacrad of the root of the diapophysis	. 2·9
Lateral diameter between outsides of the metapophyses .	. 3·7
Ditto between outsides of zygapophyses .	. 2·3
Sterno-dorsal diameter to crown of neural arch, distal end .	. 3·4

No. 125. This *lumbar*, being the one that articulates with the sacrum, differs from the corresponding one in the domestic cow in the same characters that have been noticed in the preceding two, except that the notch between its zygapophyses is little less open than in the cow. The edge of the articulation formed by these processes curves, as in the other lumbars, over the prozygapophysis of the sacrum. In the domestic cow a low acute ridge is prolonged from the metapophysis obliquely across the base of the diapophysis in this and the two preceding lumbars, but nothing like it appears in the fossils.

On the sacral edge of the neural arch, between the zygapophysis and the diapophysis of the three fossil lumbars here described, there is a small angular point, less conspicuous on the last than in the two preceding ones. A similar point appears on several of the lumbars of the musk-cow, but none can be detected in the skeleton of the domestic cow: In the musk-cow this point gives additional resistance to the joint that adjoins them, and may be considered as a minute anapophysis, but in the fossil lumbars, the points being removed from the joints and nearer the diapophyses, cannot serve such a purpose. The depression on each side of the atlantal edge of the neural spine is slight and not so well defined as in the two preceding lumbars.

Length of centrum on its sternal side, including the convexity of its antarticular surface	3·3 inches.
Ditto ditto ditto excluding the antarticulation	2·9
Ditto ditto dorsal side, excluding the antarticulation	2·8
Length from proximal edge of prozygapophysis to distal edge of zygapophysis . .	4·5
Lateral diameter of the centrum at the notches sacrad of the roots of the diapophyses .	3·4
Lateral diameter between the outsides of the metapophyses	4·6
Lateral diameter between the outsides of the zygapophyses	5·3
Sterno-dorsal diameter to crown of neural arch, distal end of centrum . . .	3·3

We have not the means of determining whether this fossil species resembled the mustush and the aurochs in the number of its lumbars (five), or the domestic ox in having six.

No. 89 is a fragment of a *sacrum*, which exactly fits the last lumbar (No. 125) described above, and is in a similar state of preservation in respect to the condition of its surface; but only the two first joints remain, the rest having been broken off. The centra have firmly coalesced, but the line of their union remains apparent.

Transverse diameter of articulating surface opposed to centrum of last lumbar . .	3·2 inches.
Sterno-dorsal diameter of ditto	1·8
Transverse diameter of zygapophysial joint, inside measure	3·4

No. 139 is the greatest part of the *right ilium*, two-thirds of the acetabulum, and part of the pubal, in a somewhat different state of preservation and of a different colour from the preceding lumbars and sacrum, but seemingly belonging to an animal of the same size. The acetabulum is scarcely less than that of a domestic ox of the largest size. As compared with the same bones of the domestic ox or of the aurochs, the most apparent difference is in the sternal edge of the ilium immediately over the acetabulum. This is acute in the fossil, whereas it is much rounded in the ox, and rounded also but less so in the aurochs. The depression situated mesiad of this edge is a deep pit in the fossil, and the shallower one on the other side of the ridge is rougher than in either of the species above named; but it has considerable resemblance to the mustush ilium, and differs from a fossil pelvis dug up at Grays on the Thames in this part.

When compared with the innominatum of the adult aurochs bull in the British Museum, it is found to be stouter and larger. An oblique ridge on the dorsum of the ilium is more evident in the fossil, and the space before it more inclined. The curve of the ischiatic notch is also somewhat different in the two, and there are other discrepancies observable on a close comparison, which are not easily described. Its size is still greater in comparison of the same parts of the skeleton of the mustush-cow, but otherwise it is more in accordance with that model than on the plan of the pelvis of the aurochs.

No. 138 is a complete *left acetabulum*, with most of the *pubal*, nearly in the same fossil condition with the sacrum (No. 89) and the lumbars (Nos. 123, 124, and 125) above noticed. The brim of the acetabulum is less rounded off than in the domestic ox, the pubal is more slender, and the brim of the pelvis formed by the pubis is decidedly an acute line in the fossil, while in the young domestic ox it is much rounded, though in an aged cow it is acute.

	NO. 138.	LARGE OX.
Transverse diameter of the acetabulum	3·0 in.	3·3 in.
Proximo-distal diameter of ditto	3·5	3·7
Circumference of the pubal where it is most slender	3·9	4·7

No. 24,576. The lower end of a *left humerus*. This is one of Captain Kellett's specimens belonging to his first collection, and has been lodged in the British Museum. It resembles the humerus of the mustush in form, but is of much larger dimensions than the bone of a cow of that species with which it was cursorily compared.

	NO. 24,576.	MUSTUSH.
Transverse diameter at the condyles	4·2	3·8
Circumference of narrowest part of shaft	9·2	—

No. 79. Condyles and fragment of the shaft of a *right humerus*, much decayed. The anconal cavity has a deep pit on its radial side, which does not exist in the same form in the humerus of the domestic ox or musk-cow, nor in the anconal cavities of Nos. 81 or 99, assigned above to *Bison priscus?*. Whether this pit be a constant character or not can only be determined by the examination of more specimens; but it exists in No. 82, which is the lower end of the humerus of a smaller individual than that to which either of the above belonged. It is designed for the reception of the crescental beak of the olecranon, and its size corresponds to the greater breadth of that process in the ulna to be hereafter described (No. 72).

No. 79 corresponds in size to radius No. 74, to be hereafter noticed.

Nos. 51, 70, 71, $\frac{72}{2}$, 73, 74, 75, 76. The portion of Captain Kellett's collection sent to Haslar Hospital contains eight radii, more or less perfect, all of them superior in width to the radius of a full-grown Lincolnshire ox, and from three-quarters of an inch to an inch wider than the same bone of the adult male aurochs in the British Museum, or than that of the mustush-cow in the same institution. As the fossil aurochs is characterized by having more slender limbs than the domestic ox or *Bos primigenius*, these relics are not referred to that species, of which no remains have been identified in Captain Kellett's collection, but rather to the *Bison crassicornis*. There are however two radii in Captain Kellett's first collection, which are smaller than those at Haslar, being only 14 inches long, and less than 2 wide in the middle of the shaft, which should probably be referred to *Bison priscus?*. I have not examined them carefully along with the others. They are numbered 24,576 in the British Museum.

Nos. $\frac{72}{1}$, $\frac{72}{2}$. These numbers refer to the most perfect *radius* in the collection, and to an *ulna* which fits it, and is undoubtedly a portion of the same skeleton. Both are of a dark chocolate co-lour, heavy, having lost little animal matter, and retaining a glossy surface, not injured by friction, and very little by decay. A view of the thenal aspect of the bone is given in Plate XV. fig. 3, and fig. 4 represents the elbow-joint, of the natural size. The radius is complete, both articulating surfaces being in an excellent state of preservation. The ulna has lost the end of the olecranon nearly at the place to which the apophysis had coalesced, and also about an inch of its length, not far from its distal extremity, which is firmly united to the radius. The coalescence of the middle of the shaft with the radius, not having been complete, has broken up; and there is, as in the allied species, an interval below the elbow-joint, where the bones do not touch, and the bony union immediately ad-joining the elbow which is formed in old individuals had not advanced far in this animal, and has yielded to the violence to which the bones have been subjected.

When compared with the same bones of other *Bovidæ*, the most tangible difference exclusive of size occurs in the outline of the articulation of the ulna with the radius within the elbow-joint (Plate XV. fig. 4). Tracing the line of junction from the lateral or ulnar side, the first section of the joint is a straight, nearly transverse line in the fossil: this portion forms a curve convex thenad in the domestic ox; and in the musk-ox it is a straight line, but trends in the direction of the centre of the elbow, so as to give, in conjunction with the succeeding line of the joint, a triangular form to the process of the ulna, of which they form two bounding lines. That process in the do-mestic ox and in the fossil has a totally different shape. The next line of the joint ascends in the fossil obliquely towards the olecranon, forming an obtuse angle with the preceding one; and the third, which constitutes one half of the breadth of the radio-ulnar joint, is a re-entering curve, or almost a very obtuse angle, with its apex turned towards the centre of the elbow. In the domestic ox this last line is a much more shallow curve, almost a straight line, which scarcely forms more than one-third of the joint. In the musk-cow again this inner or radial part is nearly a straight line, and forms more than half the axis of the radio-ulnar joint; and a deep pit for a synovial gland, impressed in the root of the olecranon, touches the mesial half of it. In the fossil the synovial pit is placed more exteriorly and also in the domestic ox, but it is not alike in the two. There are also dif-ferences in the synovial depressions of the radius. Compared with the elbow-joint of the mustush, that of the fossil greatly resembles it, but no opportunity occurred for comparing it with the inte-rior of the elbow of the aurochs. In comparative size the olecranon approaches more to that of the aurochs than to that of the mustush, in which this process is proportionally small. Some differences in the relative size of the three articular cavities of the radius for the reception of the corresponding surfaces of the humeral condyles may be detected by minute measurements, but there are none that can be readily discovered by the eye. When compared with the shaft of the radius of the domestic ox the fossil shows greater convexity on its thenal surface and inner border near the knee-joint, and con-siderably more prominence of the two protuberances near its distal joint, with a greater squareness of its outer or ulnar border. The theno-anconal diameter of the olecranon of the ulna is propor-tionally great, while its transverse diameter does not exceed that of the process in the domestic ox.

	FOSSIL 72.	FULL-SIZED LINCOLNSH. OX.
Length from proximal anconal edge of elbow-joint to most distal points of carpal joint, inner side	14·8 in.	13·1 in.
Transverse diameter of elbow-joint	4·2	3·7
Transverse diameter of the radius at its protuberance, immediately below the elbow-joint	4·8	4·0

I

	FOSSIL 72.	LINCOLNSH. OX.
Transverse diameter of the shaft near its middle, where it is narrowest .	2·6 in.	2·1 in.
Transverse diameter at the tubercles, above the carpal joint . . .	4·3	3·5
Transverse diameter of surface articulating with the carpus . .	3·9	3·1
Ancono-thenal diameter near the middle of the shaft	1·5	1·1
Ancono-thenal diameter of olecranon, immediately adjoining the joint .	4·5	3·2

No. 71 is evidently the *right radius* of the same skeleton to which the preceding one belonged. Part of the ulna still adheres to it, but the ankle-joint and part of the shaft has broken away.

No. 73 is a broken *right radius*, with a piece of the ulna attached, and similar in size and aspect to the preceding. The transverse diameter of its elbow-joint is just perceptibly smaller.

No. 70 is a fragment of a *left radius*, considerably more decayed and of a larger size. The transverse diameter of the articulating surface of its elbow measures 4·5 inches, and the shaft is wider than that of No. 72, but some of its substance has scaled off, so that I do not record its breadth.

No. 75 is a paler-coloured *left radius*, in a perfect state of preservation, with the ulna broken away, but the elbow and carpal joints entire. It is a quarter of an inch shorter than No. 72, and narrower in proportion.

No. 74 is a *right radius*, similar in colour and preservation to No. 75, and might be considered as its fellow, were it not about half an inch shorter.

No. 76 is a fragment of a *right radius*, from which the ulna has been separated and the distal end broken off. It is the smallest radius in Captain Kellett's second collection.

The transverse diameter of the elbow-joint	4·0 inches.
Width at the tubercles near that joint	4·5
Width of the shaft where least	2·5

No. $\frac{52}{2}$ and 154 are fragments of *two right radii*, split longitudinally.

The *metacarpal* brought home by Captain Beechey, and figured in Dr. Buckland's plate iii. f. 7, is of stouter dimensions than the metacarpal of a large Lincolnshire ox, and particularly in its shaft, and on that account we are disposed to range it under *Bison crassicornis*.

No. 84 is a *left femur*, nearly perfect, having suffered only some slight abrasions, with no great loss of animal substance, yet very considerably lighter than the recent femur of a Lincoln-shire ox.

	NO. 84.	LINCOLN-SHIRE OX.	AUROCHS, CUV. OSS. F.
Length from the top of the head to the most distal edge of the patellar trochlea	17·4 in.	18·0 in.	— in.
Length from the top of the head to the extreme convexity of the inner condyle	17·2	18·3	—
Length from crest of great trochanter to the extreme convexity of the outer condyle	18·4	19·7	16·5
Distance between the most elevated or proximal part of the crest of the great trochanter to the apex of the little trochanter .	6·8	6·7	—
Greatest breadth of trochanter	4·1	4·4	—
Circumference of the shaft where it is most slender . .	7·1	7·1	—
Transverse axis of head and trochanter	6·8	7·1	—
Transverse diameter of condyles	5·7	5·3	4·3

	NO. 84.	LINCOLN-SHIRE OX.	AUROCHS, CUV. OSS. F.
Antero-posterior diameter of the knee, between the inner border of the patellar trochlea and the posterior convexity of the inner condyle	6·3 in.	6·8 in.	— in.
Transverse diameter of the shaft where it is smallest . . .	2·2	1·9	—
Antero-posterior diameter of the shaft at the same place . .	2·4	2·5	—

A *femur* similar to this one in dimensions, but more defective in the trochanter and more complete in the inner edge of the rotular trochlea, forms part of Captain Beechey's collection deposited in the British Museum, and has been figured on a reduced scale by Dr. Buckland (lib. cit., pl. iii. f. 3). In consideration of their size only I have been induced to range these two femora under the head of *Bison crassicornis*, not having had an opportunity of comparing them leisurely with an authentic thigh-bone of the *Bison priscus*.

No. 84, when compared with the thigh-bone of a large Lincolnshire ox, which is longer than the thigh-bone of a Spanish ox having an equal bulk of carcase, offers the most striking difference on the division of the *linea aspera* into two diverging lines, bounding a deep rough fossa taking place nearer the condyles. The fossa separating the head of the bone from the trochanter is deeper in the fossil, and more widely bordered towards the lesser trochanter, and the great trochanter has a somewhat different line of crest from that of the domestic ox. The shaft of the bone also is somewhat thicker in the fossil, but so that while its transverse diameter, where it is least, is greater in proportion, its antero-posterior diameter is somewhat less. The obtuse ridge in front of the shaft, which abuts against the inner edge of the patellar trochlea, is narrower and more prominent in the domestic ox; and the shorter one connected with the outer edge of the trochlea, which is very apparent in the Lincolnshire ox, though nearly obsolete in the Spanish one, is obsolete also in the fossil femora. In the fossils the convexities of the condyles do not project beyond or distad of the extreme edges of the patellar trochlea, but they do so in the femora of domestic oxen, as may be observed by the above table of measurements.

No. 92 is a *right femur*, more decayed than the preceding, and measuring nearly an inch less in extreme length, but presenting no other tangible differential character. This and No. 84 are part of Captain Kellett's collection sent to Haslar Hospital.

Nos. 85 and 159 are fragments of the condyles of femora, too small to offer any discriminating characters.

No. 24,576 is the head of a femur, forming part of Captain Kellett's first collection, now in the British Museum.

No. 83 is a *left femur*, which has lost the greater trochanter and the distal articulation. Its shaft is almost an inch greater round its smallest part than No. 84, and the fossa which divides its head from the trochanter has more resemblance to the same part in the ox than in the other fossils which we have referred to *Bison crassicornis*; the front of the shaft also being more rounded than in these, but the bone is too much mutilated to give full characters.

No. 86. This *right heel-bone* differs little in its articulating surfaces from that of the domestic ox, but the projecting shaft is considerably stouter and differently formed. It is very similar to the calcaneum of the mustush, though of a greater size than that of the female skeleton in the British Museum; and it is also considerably bigger than the heel-bone of the adult male aurochs in the same institution.

I 2

No. 204 is also a *right heel-bone*, very similar to the preceding one, No. 86.

Three *metatarsals* or *cannon-bones* exist in the collection at Haslar formed by Captain Kellett; one of them in an excellent state of preservation, but having lost a considerable quantity of animal matter; the other two much decayed, the outer layers of bone having partially scaled off.

No. 78, which is the one in good condition, is manifestly stouter in the shaft than the corresponding bone of a Lincolnshire ox.

	NO. 78.	LINC. OX.
Extreme length, measured posteriorly	10·7 in.	10·6 in.
Transverse diameter of tarsal joint	2·5	2·5
Transverse diameter of shaft 3½ inches from tarsal joint	1·6	1·4
Transverse diameter of distal end, at the junction of the epiphysis . .	2·8	3·0
Ditto ditto at the extremity of the joint . . .	2·9	2·9
Antero-posterior diameter of tarsal joint	2·4	2·4
Ditto of shaft 3½ inches from tarsal joint, at the edges .	1·6	1·5
Circumference of the shaft at the same place	5·6	4·9

The length of a metatarsal of *Bison priscus*, found at Clacton, is, according to Professor Owen, 11·2 inches, and its circumference in the middle of the shaft 5·2 inches, which makes it longer and a little more slender than our fossil. A metatarsal of *Bison minor* found at Grays measured, according to the same author, 11·0 inches in length and 6·2* in circumference, which are very nearly the proportions of our fossil; while the metatarsal of a *Bos primigenius*, found at the same place, was 11·5 inches long and 6·5 in circumference. The length of the cannon-bone of the skeleton of the aurochs in the Paris Museum is, according to Cuvier (Oss. Foss. vol. iv. p. 139), 10·4 inches, the breadth of its tarsal joint 2·2, and of its distal one 2·5.

No. 78 is the *right metatarsal* bone, and when compared with its homologue in a large Lincolnshire ox, the dimensions of which are recorded above, is found to differ principally in the popliteal surface of the shaft, being convex and even swelling out in the middle and especially on the inner side of the faint mesial groove, instead of being flat as in the domestic ox. In these respects it resembles the aurochs or mustush metatarsal, and the latter more than the former. Indeed, except in size, it presents all the characters of the hind cannon-bone of the mustush very correctly.

With regard to the proximal joint, the surface that articulates with the pisiform bone is oval in the fossil, and rather more transverse than in the domestic ox, where its axis, though oblique, inclines more towards an antero-posterior position. The shaft of the fossil is undoubtedly stouter than that of the ox, and principally from its greater lateral diameter; while the groove in front is wider in the fossil and not deeper.

The fossil has suffered some injury at the junction of the distal epiphysis, either by the gnawing of animals when it projected from the frozen soil, or from violence of some other kind. There is only one other bone in the collection in which the marks of a tooth may be suspected, and these are very equivocal. The muscular attachments are more strongly marked in the fossil than in the cannon-bone of the domestic animal.

Nos. 77 and 152 are also *right metatarsals*, and as nearly as possible of the same size with No. 78.

---

* Some mistake in printing this number may have occurred, as the figure does not appear to be proportionally thick.

No. 133 and 127 are two *lumbars* of young animals, as shown by the articulating epiphyses of both ends of their centra having separated. The other parts also are much injured. From the fulness of the bone at the root of the neural spine, and the want of the depressions exhibited in that situation by lumbar No. 134, referred to *Bison priscus?*, these vertebræ would appear to belong to the *Bison crassicornis,* and to be the third or fourth numbering from the sacrum.

No. 118 is a fragment of a *cervical*, apparently the *sixth*, that is clearly of a different species from the vertebræ which in the preceding pages have been referred to *Bison priscus?* or *crassicornis*. In its colour it resembles the dentata of *Ovibos maximus* above described, but departs from the musk-ox type in the prominence of its antarticulating ball, and in the greater size of its neural canal. Its resemblance is less to the cervical of an aurochs than to one of a mustush; yet it differs from the latter in having much larger and deeper depressions on the dorsal surface of the neurapophyses, sacrad of the prozygapophyses, and on each side of the neural spine. These depressions are inconspicuous in the aurochs, distinct enough in the mustush, and very remarkable in the fossil. The centrum is so much mutilated that its true form cannot be made out. The arterial foramina are large, admitting the fore-finger.

Distance from the outside of one prozygapophysis to that of the other	4·8 inches.
Distance between the outsides of the zygapophyses	4·6
Length from the proximal edge of the prozygapophysis to the distal edge of the zygapophysis	3·7
Transverse diameter at the sinuses between the prozygapophyses and the parapophyses, outside the arterial foramina	3·7

These measurements show a breadth of articulation in the oblique processes equal to that of the cervicals referred above to *Bison crassicornis.* The other dimensions cannot be determined, from the defective state of the specimen.

The bones enumerated in the above lists are, as has been occasionally mentioned under their several numbers, in various stages of decay. Some have lost most of their animal matter, and are brittle, friable, and light; others are so fresh as scarcely to give the sensation of adhesiveness to the tongue when applied to them; and these still show more or less glossiness of surface. Some even retain specks of adherent periosteum or synovial membranes in the articular cavities. These differences we should expect to find in accordance with the varying depths at which they had been buried, and their consequent exposure in a greater or less degree to the influence of the summer thaws. The horny texture of one of the cases of the horn-core of a *Bison priscus?* is like that of a recent horn, though it is splitting into laminæ. Another of these horns is more divided into layers, which have lost their translucency and much of their tenacity. Several of the mammoth tusks also have exfoliated, and a beautiful blue phosphate of iron has formed between their plates. This is evidently the blue pigment used by the native tribes on the coasts of Beering's Sea, and which has passed from tribe to tribe by barter in small quantities as far as the banks of the Mackenzie. It is mentioned by Cook, but its origin was unknown until now. Dr. Davy had the kindness to analyse this substance at my request, and he found that the first portions I sent to him were accompanied by a greater proportion of carbonate of lime than a recent tusk should contain. The iron may have been derived from the red gravel bed, associated with the bones; but the source of the lime is not so apparent, as Dr. Goodridge observed no calcareous springs nor limestone in the soil or neighbourhood of the ice-cliffs, and the matrix of the fossils seems, both from the descriptions given of it, and a careful examination of quantities of it still adherent to the bones, to have been derived entirely from the disintegration of gneiss or granite. Having sent a second specimen of a decaying tusk to Dr. Davy, he says, " It is stained by peroxide of iron without any phosphate. I cannot find in it any mass of carbonate of lime. The proportion of animal matter remaining in it is large, sufficient to preserve the form of the fragment after the removal of the phosphate of lime by an acid. Probably complicated affinities are engaged in the production of the blue phosphate, and carbonate of iron is concerned (not carbonate of lime). Perhaps the protoxide of the carbonate may combine with the phosphoric acid of the bone, and the carbonic acid of the former with the lime of the latter; the animal matter present in the bone in clay preventing the higher degree of oxidation." Some of the bones that seem to have been lying long in the peaty soil near the surface have their cavities filled with that substance and dead fibrous roots of vegetables, together with many spiculæ of disintegrating bone. On these spiculæ there are specks of iron-glance and minute spots of ochre. The micaceous matrix in which the fossils were originally deposited, and from which many of them have been dislodged and brought towards the superficial soil, by the subsidence of the cliff through the melting of the subjacent ice, is in parts gritty from the presence of minute grains of quartz, and when dry is little coherent; but *in situ* it seems to have been tenacious enough to have received the name of " hard clay " from the officers who explored it. That which still adhered to the bones when sent to Haslar was carefully examined with the microscope, but only some broken valves of a *Cyclas*, and a fragment of a *Sesarma*, a small crustacean, both fresh-water species, were detected. No marine shells were found.

A very fine impalpable powder, in which the glittering of points of mica were observed through a microscope, was procured from the cancelli of the skulls of the mammoth and musk-ox, from recesses into which it could have been carried only by the agency of water, and not from the mere pressure of the incumbent soil. The coarser particles of the matrix were not carried so far into the interior passages of the bones.

No bones of a carnivorous animal have been discovered as yet in the cliffs by the officers who have explored them, and no marks of gnawing are shown by any of the bones, with the exception of a pair of grooves on one (No. 42, p. 15), which may have been produced by the incisors of a rodent, and some lesion of surface in No. 78 (p. 60); but even these are very problematical. Neither have any rhinoceros bones been brought home from Eschscholtz Bay. The existence of parts of the skeleton of the musk-ox in the same deposit with those of the mammoth, gives further evidence of the high arctic habitats of the latter, if further evidence were needed.

Nos. 176, 177, are the *face-bones* and part of the *calvarium* of DELPHINUS DELPHIS. Mr. Gray, who has lately made the Cetaceous animals objects of study, and executed an able monograph of the known species, had the kindness to compare the fragments with the bones of this species procured in the northern seas of Europe, and could detect no difference. These bones are evidently of considerable age, and have lost much of their animal matter, but their surfaces are almost everywhere rough, and are not stained like the undoubtedly fossil bones of the collection; and we may therefore conclude that they were not buried in the same matrix. Their canals are filled with the fibrous roots of vegetables perfectly decayed, and forming with some sand a kind of peat, much like the wrack left on the side of a fresh-water stream. Amongst this there were two or three small water-worn pebbles of granite; but the interior recesses of the bone did not contain the impalpable micaceous sand or dried mud observed in the bones of the mammoth, musk-ox, or bisons.

It would assist greatly in our reasoning respecting the past history of the ice-cliffs, if we could ascertain that the bones of marine animals were deposited side by side with those of the mammoth; but unfortunately the specimens dug out of the undisturbed cliffs were not kept separate from those gathered out of the sand-banks washed by the tides; and the general aspect of these fragments of the dolphin's skull lead us to believe that they were not buried deeply in the earth, but lay on or near the surface, exposed to occasional fresh-water floods.

No. 178. The *vertebra* of a *Cetacean*, measuring three inches in diameter and two and a half in length. This bone has evidently been bleached by the atmosphere; and as fibrous roots of growing vegetables have twined themselves into its inequalities and foramina, it is evident that it had lain for some time near the surface, within reach of the vegetation, which in that climate penetrates a very little way, since the vegetable growth is at an end for the season about the time that the summer heat has attained its greatest influence on the earth.

No. 179 is a fragment of a *Cetacean vertebra*, much decayed and mutilated. It is 3·25 inches long. This fragment, from the matters adhering to it, would appear to have been picked from the sand-bank washed by the tides.

Other bones recognized as recent when picked up were not mixed with the fossil collection.

Throughout the Arctic lands the Eskimos have carried the bones of whales, sometimes of great size, to the high grounds, out of the reach of tides; and such relics of ancient feasts may be seen in various stages of decay in the neighbourhood of deserted villages and elsewhere.

Professor Nilsson's important paper on 'The Bovine Animals of Scandinavia' having been published during my absence from England, I did not become acquainted with its existence until after the preceding sheets had passed through the press, and I can make little more use of it here than to refer the reader to it for many details respecting the living and fossil species. Among the Scandinavian fossils he recognizes the "Urox," or *Bos primigenius;* the High-necked ox, or *B. frontosus* (Nilss. K. Vetensk. Akad., 1847); the *B. longifrons* (Owen, Foss. Mam.); and the Wisent or Wisund, *B. bison,* Linn. (*B. priscus,* Owen, *fossil*). These four species were contemporaneous in Southern Scania with the rein-deer, and their remains are found in the oldest post-pliocene strata of Scandinavia. From the *B. primigenius* of modern naturalists, or the *urus* of Roman authors, the Professor believes the *large-sized lowland races of domestic oxen* to be descended; from *B. frontosus* came the *highland races of somewhat smaller growth;* and he holds it as a question to be yet decided whether the small, hornless, deer-like cattle of Norway may not be *B. longifrons* in a state of domestication. None of the domestic breeds can be traced to the bison. He lays it down as a general rule, that a tame race is always of smaller size than the wild species from which it springs, and that the largest individuals of any one species are found in the oldest of the deposits to which the range of that species extends. He remarks also that *B. primigenius* appeared in Britain at a more ancient epoch than it did in Scania, and that therefore bones of that ox dug up in the former country are bigger than those which have been found in the latter.

To afford the reader means of comparison with the dimensions of several of the fossils described in the preceding pages, the following table has been extracted from Professor Nilsson's paper :—

| | FOSSIL. | | | | RECENT. |
	UROX*.	BOS FRONT.†	BOS LONGIF.‡	BISON PRISC.§	AUROCHS‖.
Skull.					
From the anterior border of the premaxillaries to the occipital ridge	28·4 in.	— in.	16·0 in.	25·4 in.	22·4 in.
From the anterior border of the premaxillaries to the horn-cores	25·5	—	—	—	—
From the anterior border of the premaxillaries to the antinial borders of the orbits	15·4	—	—	13·0	12·0
From the occipital ridge or plane to the inial end of the nasal bones	—	—	7·2	—	10·1
From the horn-base to the orbit	6·4	5·2	3·4	3·4	3·4
Length of the horn-cores on their concave side	18·5	—	—	—	11·0
Length of the horn-cores on their convex side	26·0	—	4·0	15·0	12·5
Length of ditto in a right line	—	—	—	10·4	8·4
Height from the upper projection of the forehead to the edge of the occipital foramen	—	—	4·0	—	5·2
Breadth of the forehead between the inial edges of the basal rings of the horn-cores	9·1	8·0	5·0	14·0	11·2
Breadth of the forehead between the antinial borders of these rings	12·2	10·0	—	15·2	10·4
Breadth of the forehead between the orbits, upper parts	12·2	10·4	7·0	14·4	13·4
Breadth of the forehead between the lower parts of the orbits	11·4	—	—	—	10·5

* The museum at Lund contains two skeletons of the urox, or *Bos primigenius*, besides ten or twelve skulls. The whole length of the skeleton, whose dimensions are here recorded, was, from the nape to the tuberosities of the ischium, 9 feet, or, including the head, between 11½ and 12 feet.

† *Bos frontosus.* The measurements are those of the skeleton of an old bull (?) dug up near Saxtorp in Scania. A skull of this species exists in the British Museum.

‡ This (*Bos longifrons*) is the smallest of the ox tribe, as far as we know. To judge from the skeleton, it was 5 feet 4 inches long from the nape to the rump-bone, and, including the head, about 6 feet 8 inches. The dimensions recorded above are the usual ones of young individuals.

§ This fossil specimen of *Bos priscus* is in the Zoological Museum of Lund. It was dug up at Bjers-johlm, near Ystad, in Scania.

‖ Skeleton of an adult Lithuanian aurochs bull, presented by the Emperor of Russia to the British Museum. Professor Nilsson considers the fossil and living aurochs to be the same species, differing only in size, the more ancient being the largest. To the present habitats of the aurochs, quoted in the text from Cuvier, he adds the northern side of Mount Caucasus, where it is an object of chase with the Tscherkesser. In Europe it is now confined to the single forest of Bialowesha, in Lithuania; and it never was abundant in Scania, nor did it ever inhabit Sweden Proper.

	UROX.	BOS FRONT.	BOS LONGIF.	BISON PRISCUS.	AUROCHS.
Breadth of the forehead's narrowest part . . .	— in.	7·5 in.	5·4 in.	— in.	— in.
Chord of distance between the apertures of the ears .	12·4	—	—	—	—
Breadth of the occipital condyles	—	—	3·2	—	—
Distance between the points of the horn-cores . .	28·0	—	—	—	—
Circumference of the basal ring of the horn-core .	14·4	8·6	4·3	11·4	—
Length of frontal bones	—	12·4	—	—	—
Length of orbits	—	3·0	2·4	—	—
Length of the nasal bones	—	—	6·0	9·0	—
Length of the molar maxillary series . .	7·4	—	5·2	—	—
Length of the mandible, from tip to angle . .	20·0	—	13·2	17·4	—
Length of the mandible to the hinder edge of the condyloid process	—	—	13·5	—	—

Trunk.

	UROX.	BOS FRONT.	BOS LONGIF.	BISON PRISCUS.	AUROCHS.
Length of the spinal column to the last dorsal .	91·4	—	—	—	—
Continuation of ditto in a right line to the upper tuberosity of the ischium	9·0	—	—	—	—
Length of the seven cervicals	23·4	—	—	—	17·0
Breadth of the atlas over the wings . . .	10·2	—	4·5	8·4	8·3
Breadth of dentata	3·0	—	—	—	—
Length of neural spine of second dorsal .	14·0	—	—	19·0	17·5
Ditto ditto foremost lumbar .	—	—	—	4·2	3·5
Length of the pelvis between the tuberosities of the ilium and ischium	25·4	—	14·2	25·4	24·0
Chord of the distance between the tuberosities of the ilium	23·0	—	—	—	14·0
Length of the first rib	—	—	—	—	9·0
Length of the ninth or longest rib . . .	29.0	—	—	—	18·5
Breadth of the widest rib	2·5	—	—	2·0	—

Extremities.

	UROX.	BOS FRONT.	BOS LONGIF.	BISON PRISCUS.	AUROCHS.
Length of the scapula	20·0	—	11·4	21·0	18·4
Breadth of its base	12·0	—	6·1	12·0	—
Distance from glenoid cavity to acromion . .	—	—	1·6	3·0	—
Length of the humerus between the articulations .	14·0	—	8·5	15·4	14·0
Breadth of the distal articular surface of the humerus	—	—	2·4	—	—
Length of the radius	14·4	—	10·0	14·5	13·4
Length of the ulna with olecranon	19·6	—	—	—	—
Length of the metacarpus between the articulations .	10·0	—	7·3	8·5	7·5
Least breadth of metacarpus	—	—	—	2·2	—
Breadth of the distal articular surface of the metacarpus	—	—	2·0	—	—
Length of the femur between the articulations . .	19·0	—	11·4	20·0	18·0
Length of the tibia	17·5	—	11·4	19·0	17·0
Length of the metatarsus	11·0	—	8·3	10·5	9·5
Least breadth of the metatarsus . . .	—	—	—	1·6	—

K

	UROX.	BOS FRONT.	BOS LONGIF.	BISON PRISCUS.	AUR-OCHS.
First toe-joint	— in.	— in.	2·0 in.	— in.	— in.
Second ditto	—	—	1·2	—	—
The hoof (ungeal)	—	—	2.2	—	—

Professor Nilsson remarks, that if the measurements of the bones of *Bison priscus* be compared with those of *Bos primigenius*, it will be found that while all the other bones in the extremities are longer in the fossil aurochs, the *metatarsus* and *metacarpus* are longer and thinner in the urox, the rest of the skeleton of the latter being stouter. He also says that the fossil aurochs' skulls which he saw in England belonged to a different species, or at least to a much older form than the Scanian ones.

OVIBOS MOSCHATUS. *The Musk-ox.* (Umingmak of the Eskimos.) (Plate II.)

Osteology of a Musk Bull above four years of age.

The deciduous molars have been replaced, but have not attained quite to the level of the others, and the new lateral incisors have risen but a little above their sockets; hence, judging from the progress of dentition in the domestic ox, the age of this young bull may be stated at between four and five years.

Cranium.

The forehead is so much covered by the spongy bases of the horn-cores that its exact convexity is not at first apparent; but when looked at on the mesial line, in the furrow between the bases of the horn-cores, it forms a flat curve, which, springing more convexly from the occipital ridge, gradually becomes less so, until the arched form disappears wholly between the orbits, and from thence to the nasal bones the profile remains flat. Transversely the forehead is slightly convex, a shallow furrow running along the base of each arched, laterally projecting, orbital plate, whose arched form it augments. (Plate II. fig. 2.)

A loose swelling net-work of the frontal, formed of strongly-walled cells, coalesces with the spreading roots of the more finely spongy or porous horn-cores; and even at the early age of the individual which furnished the skeleton that we are now describing the proper tables of the frontal have separated nearly as far as the mesial plane of the skull, and are tied together by stout pillars, forming extensive sinuses, like those of the skull of an elephant, but having firmer and more massive walls. The bases of the horn-cores approach on the surface of the skull to within half an inch of the occipital ridge in one direction, and spread out in the opposite one to past the roots of the orbits; they also extend mesiad to within three-quarters of an inch of each other, while a rough thick corner raised above the skull projects iniad beyond the occipital plane; and the cores springing from them curve suddenly downwards by the sides of the skull, descending below the plane of its basilar aspect; they are compressed in their descending portion, their antero-posterior diameters much exceeding their transverse ones.

The horny sheaths are thick and ponderous, each weighing upwards of 5 lbs. avoirdupois*.

---

* The exact weight of each horn was 5 lbs. 2 oz., after some little loss in separating and drying; so that the bull has to carry 10¼ lbs. of horn exclusive of core.

They are very wide at their convex, comparatively depressed bases, but gradually assume a round form as they proceed towards the ends of their cores, after passing which they incline in their descent a little forwards, and then turn mainly outwards, with a regular graceful curve, their acute tips pointing upwards and slightly backwards.   The basal half of the horn-case is very rough, being coarsely columnar on the exterior surface and deeply sculptured interiorly, so that its thin crests and prominent ragged points fit into the furrows and cells of the cores.   There is moreover an interior swelling of the substance of the base of the horn-case forming a prominent cushion, which fills a cavity on the coronal aspect of the core.   The thickness of the substance of the case at that place is about three inches.   Towards their tips the fibrous surface gradually becomes smooth and compact, and the extreme points are black, the rest of the horn-case being dull or brownish-white.

In the old bulls the bases of the horn-cores spread still further, and the orbital places acquire much thickness and a scabrous surface.   Fragments of fossil skulls that have been exposed to friction have generally much of the friable bases of the horn cores worn away, but the smooth mesial groove in the situation of the sagittal suture remains, and from this circumstance Fischer named the fossil musk-bull skulls found in Siberia *Ovibos canaliculatus*.   If the fossil animal be however a distinct species, this character alone will not describe it, as it is equally conspicuous in the existing species.

In a *Bull* somewhat more than *a year old*, but which when killed had not ceased to follow his mother, the horn-cores have a purely lateral origin, and do not rise at all above the facial line, but springing from an almost cylindrical root immediately behind the orbits, stand out laterally with a moderate inclination basilad and antiniad, their axes forming with the mesial plane of the cranium an angle of 62°.   These cores are moreover, in themselves, concave on their facial or coronal aspect, by which they receive a uniform upward curve in the direction of their length, in addition to their general direction of outwards, basilad, and forwards.   The tips of the cores in this yearling extend further from the side of the skull laterally than any part of the massy core or its sheath in the four-year-old animal*.   In the yearling, moreover, the curve of the calvarium is considerably convex longitudinally, and more flatly but still decidedly arched transversely, the vertex of the forehead lying between the bases of the horns, or nearly midway between the centres of the orbits and the occipital ridge.   In possessing this convexity both longitudinally and transversely the forehead of the musk-bull differs greatly from the flat brow of the domestic ox, and the superior thickness of the walls of the skull adapt it for carrying an enormous weight of horn, and impart to it a vastly greater power of resistance.   (Plate IV. fig. 2.)

(Plate IV. fig. 1.)   The frontal of an *adult* but not *aged Musk-cow*, the mother of the yearling above mentioned, is perfectly flat, both longitudinally and transversely, from the parietals to opposite

* A Spanish ox, above four years old, which stood 64 inches high at the shoulders, and had a spread of 44 inches between the tips of its horns, has light hollow horn-cores, and much thinner cases, one of the latter not weighing beyond $2\frac{1}{2}$ lbs.

the fore third of the orbits, before which the bone is a little depressed as far as the transverse suture. The sagittal suture is complete between the frontals, but is obliterated between the parietals. These bones rise coronad above the frontals, and then slope down towards the occiput.

On the coronal aspect the bases of the horn-cores are flat in comparison with those of the bull, and, instead of rising into thick cellular protuberances, are formed of about four thin scabrous plates, lying closely one over the other. They approach within about an inch of each other, mesiad, and curve down over the temporal fossa and zygoma as in the male, but without the fold or roll facing the temporal fossa, shown in Plate III. The apex of the core scarcely descends below the inferior edge of the malar, instead of passing the basilar plane of the skull, which it does in the male, and they have a slight inclination forwards in their descent, with a still slighter one laterally. The horn-cases take the direction of the cores as far as these reach, beyond which they continue to descend a little basilad, but mainly antiniad, till they come on a line with the fore edge of the second true molar, and then they trend out laterally with a regular curve, having its concavity and the tip of the horn directed iniad. Convex at the base, and more or less compressed until it passes the end of its core, the horn-case then becomes round, tapers, and ends acutely: the base is irregularly and coarsely grooved; the fibrous structure becomes smoother and gradually finer, until it is lost in the smooth and polished tip.

Resuming the description of the *cranium* of the *four-year-old Bull*, we observe that the frontal bones are prolonged further antiniad than in the domestic ox, and consequently still more so than in the aurochs or fossil *Bison priscus**. The transverse suture is anterior to the front walls of the orbit, and is reflexed in each half of the frontal, forming a curved notch into which the end of the nasal is received, as in the sheep. In the domestic ox the nasal process of each half of the frontal forms with its fellow a single acute notch, into which the pointed ends of the nasal fit in a manner very unlike the zigzag course of the transverse suture of the musk-ox, as represented in three figures of Plate IV. In the reduced front view of the skull in Plate II. the face is viewed obliquely, and is so much foreshortened that the relative position of the brim of the orbit to the transverse suture is not clearly shown.

(Plate III.) The occipital ridge is situated half an inch posterior to the occipital suture, and when regarded from behind appears as if supported by a central buttress, resulting from its union with the occipital spine. The edge of the ridge describes a shallow concave curve on each side of the buttress, which overhangs or projects somewhat iniad of the vertical occipital plate†.

In the domestic ox the crown of the skull is situated in the occipital ridge; but in the four-year-old musk-bull it is two inches and a half further forward, being placed between the posterior thirds of the horn-cores. The distance in the musk-bull, as measured by callipers, from the crown to the

---

* In these the ends of the nasals reach as far iniad as the orbital angle of the lachrymal bone.

† In the domestic ox the occipital spine is a mere roughened spot in the concave occipital plate, which is boldly limited above by the obtuse prominent ridge that runs transversely between the horn-cores, and scarcely so far back as their posterior surfaces. This is in the large Spanish or Lincolnshire oxen. In the Alderney cow, which is perhaps as artificial as any of the breeds cultivated by graziers, the occipital crest has a different form, being puckered up in the middle so as to show a blunt mesial cone on the facial aspect and a crescentic curve on the occipital one. The skull however retains its true taurine character in regard to the backward position of the horns in respect to the plane of the occiput.

upper border of the occipital foramen, is 5·2 inches, and from the occipital crest to the same opening is 3·7 inches, or scarcely inferior to the same space in the largest-sized domestic oxen.

The occiput, including the super-, par-, and ex-occipitals, is more nearly square in the musk-bull than in the domestic ox. Its greatest width is at the auditory openings, where, taking in the projections formed by the paroccipitals and the edges of the squamosals, it measures transversely 6·7 inches; but on a level with the occipital spine its breadth is only 4·7 inches. The structure of the occipital bone is, like that of the rest of the calvarium, much stronger in the musk-bull than in the domestic animal, particularly towards the condyles, which are joined in a more convex form by the plates of the exoccipitals, and do not consequently stand out so abruptly from the general occipital surface. The condyles themselves are also more spherical in the musk-bull, and the transverse line which marks the meeting of their anterior and posterior ellipsoids is less conspicuous than in the domestic ox, in which the two articulating surfaces of the condyle unite in a visible edge. The vertical diameter of a condyle is greater than in the domestic ox, and the transverse one less. There is an additional and still more perceptible difference of form in the condyle of the musk-bull, being furnished with an exterior heel that occupies much of the space between the condyle proper and the paroccipital spine, and furnishes a pulley or trochlea, which moves on a concave, pretty broad articular surface, formed by a lateral notch in the brim of the atlas. (Plate V. fig. 1, 3.) In the domestic ox the site of this notch is occupied by an acute-edged, somewhat incurved, semi-oval lobe, which moves in the space between the condyle and occipital spine. The supplementary trochlea, and the proportionally greater width of the occipital foramen, which sets the condyles further apart, gives a more extended fulcrum to the head of the musk-bull, and adds security to the entire joint.

The basi-occipital is flat, nearly twice as wide as in the domestic ox, and in place of a ridge-like protuberance on each side, where it unites with the basi-occipital, it has an oval rough surface. It has also a mesial hypapophysial ridge and a pair of shoulders which abut against the confluent pair of notches in the centrum of the atlas on its sternal side. It is coalescent with the basi-sphenoid, which is narrow, but flat also, and is not furnished with the two crest-like projections that exist in the domestic ox, and intercept a deep mesial furrow. In the musk-bull, and also in the adult musk-cow, there is a faint indication of a raised hypapophysial line; but in the yearling musk-bull this is not perceptible, and there is even a pretty deep smooth mesial furrow in the basi-occipital separating the two rough spots.

The paroccipital spines descend straight in the same plane with the occiput, and, instead of curving mesiad, as in the domestic ox, have merely a shoulder on the inner side of their tips, which is ossified and coalescent in the adult cow, but remained in its cartilaginous condition in the four- to five-year-old bull, and, having separated during maceration, is not shown in Plate III.

The thickness of the walls of the cranium is, as has been mentioned above, very great. At the crown, between the horn-cores, it is 2·6 inches in this bull, and, including the horn-core bases, much greater.

Face.

The face tapers more towards the tips of the premaxillaries than in the ox, the muzzle being narrower, which gives that portion of the skull an ovine aspect; and the orbits project fully two inches and a half from the lateral planes of the squamosals and maxillaries; indeed the position of the horns renders it necessary that the eyeballs should be stilted out to enable the animal to see in any other direction than straightforwards. The horn applies so closely to the zygoma behind the orbit that there scarcely appears to have been space for the interposition of integuments.

The lachrymal differs considerably in form from that of either the ox or sheep, being broader anteriorly, and having its axis lying more parallel to the mesial plane of the skull than in the ox. In its connections and relative size this bone varies with the age of the musk-ox. In the fœtal skull (Plate IV. fig. 3) it is triangular; one of its corners just touches the point where the outer angle of the nasal process of the frontal, the lateral corner of the nasal proper, and an angle of the maxillary meet, the lachrymal not being continued along the side of the nasal at all. In the yearling (Plate IV. fig. 2) the lachrymal and nasal touch each other through a little greater space, and in the adult cow (Plate IV. fig. 2) this space is still further augmented. There is however some little variation in the extent on different sides of the face in the same individual, there being rarely perfect symmetry in these parts.

The nasals end in a tapering point anteriorly, without the narrow and generally oblique notch which exists in the tips of these bones in the ox and aurochs, though there is generally a minute indentation further back and near the maxillaries. The nasal of the sheep has the same kind of tapering point, without a notch, with that of the musk-ox; but the notches at the end of the nasals are not very constant in form in the same species. In some domestic oxen they are deep, with equal bounding processes, resembling those of the buffalo; and in an aged Alderney cow the notch is deep and narrow, with its lateral corner prolonged beyond its mesial one.

In the common ox the inial end of the premaxillary reaches the side of the nasal; in the musk-bull a portion of the edge of the maxillary intervenes between them*. This is also the case in the aurochs. The opening between the branches of the premaxillary is narrower than in the ox, and rather longer in proportion, and the palate has a different form in the two species, not being convex or arched in the musk-bull of four or five years, but level along the mesial line, and as wide as in the largest Spanish or Lincolnshire oxen. It is however more concave transversely, and narrows anteriorly by about one-third of its breadth, the line of the molars curving in that direction towards the mesial plane. In the domestic ox the interior sides of the molars on the right are parallel to those on the left, except the foremost pair, which stand rather more inwardly. The notch between the plates of the palatines is rounder and wider in the musk-bull than in the domestic animal.

The mandible of the musk-bull is less arched in its under outline than that of the common ox, the premolar part having considerably less rise; moreover the symphysis is shorter and more vertical, and it forms an angle or chin. The space occupied by the molars is much more than one-third of the entire length of the mandible, measured from the crowns of the incisors to the posterior edge of the ascending branch, and exceeds the space between the foremost molar and the edges of the incisors. In the large Spanish and Lincolnshire oxen the space filled by the molars scarcely exceeds one-third of the length of the whole mandible with the incisors, and is not quite equal to the part before the first premolar.

Teeth.

When the upper teeth of the mature musk-cow are compared with those of a large Spanish ox their general form appears to be much alike, and their transverse diameters to be nearly equal. The exterior lateral folds however are more prominent and more sharply defined in the musk-cow, while the

---

* In the mature fœtal skull of the common ox the premaxillary runs along the side of the nasal and touches the lachrymal, but soon after birth a part of the edge of the maxillary interposes between these bones. In the rein-deer an exterior plate of the turbinal, that coalesces in the third or fourth year with the maxillary, interposes between the end of the premaxillary and the nasal.

inner surfaces of the last three, or true and double-lobed ones, have more rounded lobes, without the intermediate, narrow, shorter pillar or fold which exists in the common ox, aurochs, and mustush. In the yearling musk-bull there are only five upper molars, all but the anterior one being bi-lobed. Of these three are deciduous, and in the four-year-old bull their place is occupied by single-lobed premolars.

The mandibular molars of the adult musk-cow, six in number, present each of them some difference from the corresponding tooth in the ox, and the three posterior or true ones want the accessory fold of enamel which occupies the bottom of the sinus that exists between the outer lobes in the domestic ox and bison. The mandibular molars, it must be observed, have their sides as it were reversed when compared with the maxillary ones, the round lobes being on the outside. The crowns are simpler, the interior folds of enamel not being so much inflexed as in the domestic ox, and the true mandibular molars in the latter have each a small interior island or ring within the crescentic fold, that does not exist in the musk-cow. The posterior lobule of the ultimate molar is small in comparison to that of the ox. In the yearling musk-bull there are five mandibular molars, with a sixth just protruding from the socket behind the others. The third deciduous one is three-lobed, and the two true molars two-lobed. In a domestic calf about two months old only three mandibular molars have protruded beyond their bony sockets, but a fourth is just showing its point. The third of these is three-lobed, with two shorter additional folds on the exterior side.

The mandibular molars and premolars of the four-year-old bull or adult musk-cow occupy a length of jaw equal to that which these teeth do in a large Spanish ox, but the incisors are much smaller and have less spread. The incisors also wear away differently, being worn down on their antinial surfaces in the three musk-ox skulls of different ages here commented on.

Skull of a fœtal Musk-ox Calf (supposed to be six months old). (Plate IV. fig. 3, 4.)

The frontal meets each nasal by a zigzag line, with a single indentation instead of the sloping suture that exists in the common ox. The sagittal suture extends backwards to the vertex of the skull, where it is crossed by the coronal suture, and at the point of intersection a Wormian bone fills the space that is occupied by a membranous bregma in a fœtal domestic calf near maturity. In this latter the apex of the convexity of the skull is situated before the bregma, and consequently nearer the orbits than in the musk-calf.

The parietals differ greatly in shape in the two species. In the musk-calf the coronal suture runs almost directly across, and the sagittal suture is continued towards the occipital spine for more than half the length of the parietals, each of which has on the whole the form of a spherical triangle, with the space between their apices and the occipital suture occupied by four Wormian bones, arranged so as to form a triangular interparietal. In the fœtal calf of the domestic cow the parietals have an irregular lunate form, and are separated completely from one another by a large triangular interparietal, whose apex reaches the bregma. In the aurochs, according to Bojanus, the interparietals are square.

The superoccipital differs little from that of the calf of the domestic animal. The paroccipitals are conspicuous, but have not that lateral prominence margined by the reverted edge of the squamosal which is exhibited even in the yearling, and forms a striking character in the four-year-old bull. In the adult cow the upper portion of the paroccipitals is less wide than in the bull, and the bone has partially coalesced with the super- and ex-occipitals.

The round form of the calvarium, and other peculiarities, are exhibited in the figures. The sockets of the maxillary contain two bi-lobed molars on each side, and an anterior one of a single lobe not quite advanced to a level with the edge of the bone.

The following measurements and all the others in this paper, unless otherwise specified, are ascertained by callipers, and of course give the chords between the points named :—

	MUSK-BULL, 4–5 YEARS.	ADULT MUSK-COW.	SPANISH OX.
Length of skull from antinial end of premaxillary to occipital ridge	20·2 in.	18·4 in.	19·9 in.
Length from same point to transverse suture at the root of the nasals, mesial line	9·9	9·6	11·0
Length from transverse suture to the occipital ridge	10·6	9·3	9·2
Breadth at the orbits, coronal edges	8·5	7·3	6·9
Ditto basal edges	9·9	8·9	8·8
Breadth immediately behind the projection of the orbits	5·4	5·0	7·1
Breadth at the reverted occipital edges of the squamosals, overlapping the paroccipitals	6·7	6·0	9·4
Distance from lateral edge of one condyle to that of the other on the occipital aspect	4·4	4·5	4·3
From occipital suture to basilar edge of the foramen magnum	5·0	4·5	5·6
From occipital suture to basilar edge of occipital ridge, mesial plane	0·6	0·6	0·9
From basilar edge of occipital foramen to antinial end of basi-sphenoid	4·2	4·0	4·1
Breadth of shoulders of basi-occipital	2·3	2·3	2·4
Breadth of the middle of the basi-occipital	2·0	2·0	1·1
Breadth of the basi-sphenoid at its antinial end	0·7	0·7	0·6
Greatest width of the opening to the posterior nostrils	1·5	1·6	1·0
Chord of the space occupied by the series of six maxillary molars	5·5	5·5	5·4
Distance from foremost premolar to tip of premaxillary	5·0	5·0	5·7
Breadth of palate between foremost molars	2·3	2·2	3·1
Breadth of ditto between the fourth molars	3·1	3·2	3·5
Breadth of ditto between second lobes of the sixth molars	3·3	3·2	3·4
From the outside of one maxilla, at the fangs of the fourth molar, to the same part of the other*	5·4	5·2	6·2
From roots of the incisors to posterior border of condyle	14·6	15·2	15·9
From ditto to posterior curve above the angle of the mandible	14·2	14·6	15·6
Rise of the coronoid process above the surface of the condyle	1·4	1·4	2·1
From last molar to posterior curve above the angle of the mandible	4·3	4·6	5·2†

Cervicals.

The cervicals are as usual seven in number, and in the four-year-old musk-bull they measure along their sternal surfaces, including the full spaces occupied by the intervertebral substances, 16

* This point of the maxillaries forms a defined protuberance in the domestic ox; but it is merely an irregular bulging, and a slightly raised line in the musk-bull.

† See page 25 for other dimensions.

inches; when the neck is raised as in Plate II., the chord formed between the occipital condyles and the neural spine of the seventh cervical is 10 inches long. These cervicals differ from those of *Bos* or *Bison* in the shortness of their centra, their fullness and flatness on the sternal aspect, the want of the hypapophysial ridge shown by the four intermediate cervicals in the other genera, and in the smaller obliquity of the brim of their distal articulating cups.

The *atlas* (Plate V. fig. 1, 2, 3, 4) is much more strongly made than that of the domestic ox, and it differs remarkably from the atlas of that species and of the aurochs in the form of its articulations with the cranium. The extreme width of the cup in which the occipital condyles move is equal to that of the large Spanish ox. In the domestic animal the prozygapophysial section of the cup meets the rounded edge of the centrum in a comparatively shallow and rather acute lateral notch. In the atlas of the musk-bull the brims of both these portions are as it were truncated, by which a wide notch is formed, as represented in fig. 1 and 3. At the bottom of the notch a deep pit is sunk, chiefly on the prozygapophysial side, apparently for lodging a synovial gland, and a comparatively roughish band extends from it across to the neural canal, marking a portion of the cup that is protected from the full pressure of the condyles. The rest of the lateral notch is polished. In the atlas of the mustush there is an approach to the lateral notch here noticed, with its accessorial articulating facet, but this is less perfect than in the musk-bull. Though the mustush is moderately horned, its head is large and heavy, and this additional security to the joint may be required to support the weight. In the structure of the articulation between the skull and atlas the aurochs has more resemblance to the common ox. The coincidence of form in this important joint, between the mustush and musk-ox, furnishes us with a typical character for the American or Arctic *Bovidæ*.

In front of the centrum is a pair of notches which receive the shoulders of the basi-occipital; and are separated from one another by a mesial groove with raised edges, and from the lateral notch by a strong, elevated, narrow, yet truncated process. In the domestic ox the floor of these notches joins that for receiving the condyles imperceptibly,—the two forming a convex surface; but in the musk-ox the two surfaces are on decidedly different planes, that of the notches being inclined much sternad. The atlas of the domestic ox differs also, in having a low ridge instead of a groove between the notches, and the rougher surface, between the zygapophysial articulating facet and that of the centrum, reaches from the neural canal only half-way to the lateral edge of the cup, the more acute-edged and laterally tapering condyle of the common ox enabling it to play on the bottom of the cup there.

The distal articulating surface is more concave in the atlas of the musk-bull than in that of the domestic ox, in which it is more undulating, and slopes less regularly towards the neural canal. A hypapophysial knob or abbreviated spine is equally prominent with that of the ox, but rises less abruptly from the centrum, and it dips a little sacrad of the articulating cup, there being a depression in the centrum of the dentata for its reception. The dorsal protuberance representing the neural spine is broader than that of the ox, and has less tendency to form a longitudinal ridge; a wider and shallower notch separates the prozygapophyses, and the space between the arterial foramina is greater on the dorsal surface: on the sternal side the distance between the foramina is the same in the four-year-old bull as it is in the large Spanish ox.

The wing-like pleurapophysial process of the atlas is stouter in the musk-bull, and has a much less distinct and extensive depression on its sternal aspect, beneath the arterial foramen, than in the common ox; its posterior or sacral angle ends in the same plane with the postarticular surface,

L

while in the ox it droops more than an inch beyond it.   The autogenous margins of these wings had not coalesced with the bony part in our specimen, and have separated in maceration, so that a line or two would require to be added to the breadth of the atlas in the figure to represent its size when complete.

Within the neural canal, on the side of the centrum, there are rough protuberances separated by a mesial uneven groove, of which there is no vestige in the common ox.   A small glazed surface on each of these ledges shows that the edge of the odontoid process plays against them.   On the lateral side of each protuberance there is a depression situated under the interior pair of arterial foramina. A more marked depression occurs in the same situation in the domestic ox, but the space between them is merely slightly uneven, not tubercular.   In the fossil atlas referred to *B. crassicornis* (No. 90, Plate XII. fig. 4, page 43) a groove runs transversely from one depression or pit to the other.

The fossil musk-ox atlas, No. $\frac{126}{2}$ (page 22), corresponds in the dimensions of its principal articulating surfaces, neural canal, etc., with the one described above, but it has smaller mesial knobs on both the sternal and dorsal aspects, and the neural knob is divided by a longitudinal furrow, of which there is a slighter trace in the recent atlas.   The fossil is probably a relic of an aged musk-cow not bigger than the existing species.   Its lateral processes have been broken away.

	MUSK-BULL.	LINC. OX.
Extreme breadth between the most lateral parts of the atlas, distal end*	7·2 in.	8·1 in.
Breadth above the proximal end of the lateral processes	5·0	5·4
Distance between the outer borders of the zygapophysial articulating surfaces	4·3	4·6
Ditto, ditto, including breadth of articulating surfaces of the lateral notch in the musk-bull	4·8	—
Distance between the peripheral points of the hypapophysial and neural spines	3·8	3·9
Lateral axis of the neural canal, proximal end	1·5	2·0
Sterno-dorsal ditto, distal end	1·7	2·0
Breadth of the distal articulating surface	4·3	4·6
Distance from the proximal edge of prozygapophysis to sacral surface of lateral processes	3·5	5·3
Length of centrum, mesial plane, excluding the hypapophysial knob	1·8	1·8

(Plate V. fig. 1, 5 ; Plate XI. fig. 1.)   The *dentata* or *second cervical* differs more remarkably from its homologue in other bovine animals than any other segment of the vertebral column, the caudal ones alone excepted.   It is a compact, massive, strongly-formed bone, with little of that hour-glass or corset-like form so characteristic of the dentata of the common ox, when regarded on its sternal aspect.   The odontoid process is shorter, with thicker walls and a more obtuse edge.   The articu-lating surfaces on each side of the process are also differently shaped from those of the common ox, being considerably narrower in their sterno-dorsal diameter : in the atlas of the domestic ox they embrace the odontoid, covering the entire atlantal surface of the centrum, except a narrow mesial notch on the sternal side.   In the musk-bull, instead of this mesial notch, there is a broad rough surface, partly depressed, to give space and a rest for the hypapophysial knob of the atlas ; and the smooth surface is also restricted towards the sides of the neural canal by a rough spot with a groove in its middle, probably for lodging a lubricating gland (Plate V. fig. 5).   Laterally the proximal margin of the centrum flanges out a little ; but this is not the case on the sternal side ; while in the

* In the musk-bull, minus the narrow autogenous epiphyses.

common ox it projects on every side except in the notch from which the thin hypapophysial ridge descends. The whole proximal surface of the centrum is more convex, both transversely and sterno-dorsad, than in the ox, in which it is flat, with a shelving off near the edges, showing that in the musk-bull there is more freedom in nutation and direct lateral motion, but less extensive rotation than in the common ox.

The crown of the neural arch is constructed more like the segment of a circle than in the ox, in which it is a flatter ellipsis. The neural spine is stouter throughout and thicker at its proximal corner than elsewhere, and it is proportionally higher there than in the ox, in whose dentata the enlarged knob of the spine is situated at its distal corner.

The diapophysis is a long, conical, obtuse process, not perfectly round, but showing obscurely a sternal and dorsal edge: it is very different from the more expanded thin process in the domestic ox, with its ragged edge, incurved sternad and mesiad. The root of the process is of the same extent in both species; but the point in the musk-ox extends further sacral, overlapping half the centrum of the third vertebra, and passing dorsad of its diapophysis. In both species the tip of the process is edged by a narrow epiphysis, which does not coalesce till the animal is about five years old. Neither have a distinct parapophysis.

The hypapophysial ridge, which is thin, prominent, with a projecting angle near its atlantal end, and a peaked distal point in the common ox, is altogether wanting in the musk-bull, the sides of the centrum being full and rounded; but there is a broad prominence on the meso-sternal margin of the distal articulating cup, representing the peaked process of the ox. In both species this process or prominence is formed chiefly by the autogenous articulating cup, which in the musk-bull has nearly coalesced with the centrum. The abbreviation of this part in the musk-bull renders the distal articulating cup less concave and less oblique than in the common ox.

There are no distinct prozygapophyses in the dentata of either the musk-ox or domestic species, unless these processes be represented by the lateral parts of the odontoid. The zygapophyses are distant from each other and present smaller articulating surfaces in the musk-bull than in the ox, and the neurapophyses which support them are much thicker in the former.

	MUSK-BULL.	LINC. OX.
Breadth of the antarticulating surface	4·3 in.	4·4 in.
Breadth of the postarticulating cup	2·2	1·9
Sterno-dorsal diameter of ditto	1·9	2·6
Height of odontoid on the sternal side	0·7	1·0
Length of centrum, sternal side, excluding the articulations	2·2	5·0
Its width just atlantad of the root of the diapophysis	3·6	2·3
Distance from sternal side of centrum, atlantal edge, to the atlantal corner of neural spine	3·7	3·5
Distance from sternal side of centrum, sacral edge, to the sacral end of neural spine	5·4	6·2
Distance from the outer side of one zygapophysis to that of the other	2·5	3·1
Length of neural spine, peripheral edge	3·0	4·3
Length of neural arch or base of spine	2·4	3·5

Many of the differences between the dentata of the musk-bull and the fossil one, No. $\frac{90}{2}$, of *Ovibos maximus*, have been already mentioned (p. 26). To these we may add here, that though the fossil is in most dimensions larger, the length of its neural arch is a little shorter than that of the recent one.

The *third, fourth,* and *fifth* cervicals differ from those of the ox in the shortness of their centra, the want of a hypapophysial ridge, the flatness of their sternal aspect, in the absence of the peaked corset-form, and in the comparative flatness of the articulations of the centra. The para-pleurapophyses and the diapophyses have also very different forms in the two species. These processes are stouter in the musk-bull, and there is little difference in their shape in the three cervicals enumerated, except that the notch between them is deeper in the fourth, and still deeper in the fifth, than in the third one. In the domestic ox there is a more perceptible change of form in the way that the parapophyses, which in the third cervical lie in the same oblique line with the diapophyses, begin to separate from them in the fourth, are more distinct in the fifth, and wholly apart in the sixth. Moreover the parapophyses of the ox point atlantad and rise in that direction beyond the articulating ball, while in the musk-ox they have scarcely any atlantal inclination, but point diagonally between the sternal and lateral aspects, the diapophyses standing laterally and a little sacrad.

There is also a difference in the neural canal, which in these three and the succeeding cervicals approaches to a pointed arch, with its crown rising dorsad of the articulating surfaces of the oblique process. In the common ox the arch is flatter, and scarcely rises dorsad of any part of these surfaces.

In the third and fourth cervicals of the musk-bull a distinct metapophysis rises upwards of a quarter of an inch atlantad of the articulating facet of the prozygapophysis. This process is smaller in the fifth cervical, but it forms a distinct ridge in the sixth and becomes a little bigger again in the seventh. In the domestic ox a vestige of a metapophysis may be observed in the third cervical, but it is wholly absent in the others.

	CERVICALS OF MUSK-BULL.		
	THIRD.	FOURTH.	FIFTH.
Breadth of the antarticular surface of the centrum	1·9 in.	1·9 in.	2·0 in.
Sterno-dorsal diameter of ditto	1·8	1·8	1·8
Breadth of the postarticular cup of the centrum . . .	2·2	2·3	2·4
Sterno-dorsal diameter of ditto	2·1	2·1	2·0
Length of centrum, sternal aspect, excluding articulations .	2·0	1·9	1·8
Length from sternal surface of centrum and summit of neural spine .	5·1	5·3	5·8
Breadth from tip of one diapophysis to tip of the other, minus the thin epiphyses	5·8	5·7	5·3
Distance between outside corners of the parapophyses, also minus the epiphyses	4·5	5·0	4·7
Distance from the most lateral part of one prozygapophysis to that of the other	2·9	3·0	3·0
Distance from the most lateral part of one zygapophysis to that of the other	2·6	2·5	2·6
Length of neural spine above the crown of the arch, proximal edge .	2·1	2·5	3·0
Transverse distance between the outer wall of one arterial foramen and that of the other	2·4	2·5	2·7

The *sixth cervical* has no hypapophysial ridge in either the musk-ox or domestic bull, but in the latter the space between its parapophyses is narrower and less flat than in the former, and the line of junction of the articular epiphysis with the centrum differs in the two. In the musk-bull the distal epiphysis is morticed into a wide flat notch, and the proximal one is also sunk at the edges, within an uneven ledge of the centrum, which exists, though in a less degree, in the three preceding cervicals, and is still more developed in the seventh. In the domestic ox, the much more convex and generally semi-oval articulating ball rather overhangs the line of union with the centrum, instead of being received into the latter.

The parapophyses diverge more in the musk-bull, and terminate evenly by a pleurapophysial epiphysis, which is nearly of the same width and thickness throughout: in the domestic ox these epiphyses have a triangular expansion at their distal corners. The parapophyses are shorter and the neural spine longer than in the domestic ox. The metapophysis is less conspicuous in the sixth than in any of the three immediately preceding cervicals, or than in the one which follows. The cervicals of the domestic ox show no prominence in the situation of the metapophysis of the musk-bull, the third alone exhibiting a minute projection of the back of the zygapophysis above the articulating surface.

	MUSK-BULL.	LINC. OX.
Breadth of the antarticular surface of the centrum	2·1 in.	1·5 in.
Sterno-dorsal diameter of ditto	1·9	2·0
Breadth of the postarticular cup of the centrum	2·2	2·0
Sterno-dorsal diameter of ditto	1·9	2·3
Length of the centrum, sternal aspect, excluding articulations	1·4	2·4
Distance from the tip of one diapophysis to the tip of the other, minus the epiphyses in the musk-bull, but including them in the ox	4·7	6·2
Breadth between the outer proximal corners of the parapophyses, including epiphyses	4·9	4·3
Distance between the most lateral part of one prozygapophysis and that of the other	3·0	3·9
Distance between the most lateral part of one zygapophysis and that of the other	3·9	4·2
Length of neural spine above the crown of the arch, proximal edge, minus the epiphyses	3·3	3·7
Length of neural spine above the crown of the arch, distal edge, minus the epiphyses	3·5	3·3
Transverse distance between the sinuses, formed by the prozygapophyses and diapophyses, outside of the arterial foramina	3·0	2·9

Plate XI. fig. 5 represents the distal aspect of the *seventh cervical* of the musk-bull, drawn of the natural size. In this, the convex antarticular epiphysis is morticed more deeply and squarely into the centrum on the sternal side than in the preceding cervicals. It has a broader and flatter sternal front than its homologue in the domestic ox, and both want the hypapophysial ridge. There are no parapophyses, unless the half cups for the reception of the heads of the first pair of ribs be their representatives. The diapophyses are shorter, simpler, and less rough at the ends than in the domestic ox, and the neural spine tapers, while in the common ox it carries its atlanto-sacral diameter to its rounded end. Owing to the elevation of the more pointed neural arch, it rises dorsad of the apophysial articulating surface; a smooth facet, as represented in the plate, is reverted on the outer side of the arch for articulation with a corresponding facet of the first dorsal, thus giving to the joint some resemblance to those of the lumbars. In the domestic ox the flat and rather concave articulating surfaces of the zygapophyses are nearly on a level with the crown of the depressed elliptic arch, and have no reverted facets.

	MUSK-BULL.	LINC. OX.
Breadth of the antarticular surface of the centrum	2·1 in.	1·6 in.
Sterno-dorsal diameter of ditto	1·9	2·1

	MUSK-BULL.	LINC. OX.
Breadth of the postarticular cup of the centrum, including cups for heads of the ribs	3·0 in.	3·2 in.
Sterno-dorsal diameter of ditto	1·4	2·3
Length of the centrum, sternal aspect, excluding articulations	1·4	2·4
Distance between the tip of one diapophysis and the tip of the other, including their epiphyses	4·9	6·4
Distance between the most lateral part of one prozygapophysis and that of the other	3·1	4·2
Distance between the most lateral part of one zygapophysis and that of the other	2·7	3·5
Length of the neural spine above the crown of the arch, proximal edge, minus the minute epiphyses	4·8	4·3
Length of ditto, distal edge, minus the epiphyses	4·8	4·2
Transverse distance between the sinuses, formed by the prozygapophyses and diapophyses	3·2	3·3

Dorsals.

The musk-ox has *thirteen dorsals*, whose centra are more cylindrical than those of the domestic ox, and are destitute of the hypapophysial edge or ridge produced in the latter by the concavity or pinching in of their sides. In our skeleton of the musk-bull, the intervertebral substances have shrunk and drawn the vertebræ closer together than in the recent state; in this condition the dorsals measure over their spines 22 inches, and about 25 over the sternal surface of the centra. Of the series the *first dorsal* has the broadest and flattest centrum, presenting on its sternal aspect a very flat elliptical curve transversely, the fulness being carried out to the sides. Its transverse diameter is equal to its atlanto-sacral one, and exceeds its sterno-dorsal one. The breadth of the centrum diminishes gradually in the succeeding dorsals down to the ninth, and then increases again progressively to the last lumbar, which has the widest centrum of all.

The cup on the diapophysis for articulating with the tubercle on the shoulder of the rib becomes smaller on each succeeding dorsal down to the eleventh, whose rib is furnished with a distinct tubercle. Cup and tubercle are wanting in the twelfth and thirteenth dorsals and their corresponding ribs, the latter having merely some roughness at the angle. In the domestic (Alderney) aged cow all the thirteen ribs possess well-formed tubercles, which articulate with the diapophyses.

In the eleventh dorsal of the musk-bull, a protuberance of the diapophysis, resembling an anapophysis, projects sacrad of the cup for receiving the pleurapophysial tubercle. This projection is much larger on the twelfth and thirteenth, which have no such cups, and in the latter it shows its true nature in approaching the form of the para-pleurapophyses of the lumbars, but differs from them in its direction being a little sacrad as well as laterally. These two dorsals have also neural spines much resembling the lumbar ones, so that their general character as vertebræ is intermediate between that of the anterior dorsals and the lumbars.

In the *first dorsal* of the musk-bull the articulating surface of the prozygapophysis has a sigmoid flexure transversely, a portion of it running dorsad on the neurapophysis to articulate with the interior facet of the seventh cervical represented in Plate XI. fig. 5. In the corresponding dorsal of the domestic ox the prozygapophyses have simply convex articular surfaces, which do not encroach at all on the backs of the neurapophyses. The fossil first dorsal, No. 121, described in p. 51, resembles the common ox in this part, with, however, flatter articulating surfaces and less of the neural arch intercepted between them. In the aged Alderney cow the zygapophysial joint of the twelfth dorsal is the first that exhibits the change to the lumbar type of articulation, and all its thirteen ribs have an articulating tubercle, with a corresponding trochlea on the diapophysis, which is stilted out as it were in the posterior dorsals. In the twelfth dorsal a small point projects sacrad of the pedicellated diapophysial trochlea; there is an angle in the same situation in the thirteenth dorsal, and a thin corner on the sacral edge of the diapophysis of the first lumbar. True anapophyses are feebly and rarely developed in the ruminants, according to Professor Owen. The thirteenth neural dorsal spine is the first that assumes the lumbar form in the domestic cow.

There is a rough spot over the large diapophysial trochlea for receiving the tubercle of the first rib; and a distinct blunt conical metapophysis on the second dorsal projects atlantad beyond the edge of the prozygapophysis, close to the concavity which articulates with the tubercle of the second rib. It is of nearly the same size in the succeeding dorsals, with some variations of form, being in some blunt and irregular and in others more conical, down to the eleventh dorsal, in which it is the longest of all, and has a conico-subulate shape. In each successive dorsal the metapophysis has, by almost imperceptible degrees, increased its distance from the diapophysial articulating cup, until in the eleventh it forms a projection atlantad from the sternal side of the protuberance on the back of the prozygapophysis. In the twelfth and thirteenth dorsals the zygapophysial articulation has become lateral instead of dorsal in its aspect, and the metapophysis changed to a mere smooth half-pear-shaped swelling on the back of the prozygapophysis, without any projecting points. It retains this form in the anterior lumbars, acquiring more prominence in the third and fourth, and standing as a blunt ridge on the outside of the prozygapophysis in the fifth, but in the sixth or last it subsides again into a rounded tubercle. In the adult musk-cow the metapophysis of the fifth lumbar has more the tubercular than ridge form, and the smooth tubercle on the sixth is more prominent than in the musk-bull.

In the mounted skeleton of the musk-bull (Plate II.) the fourth dorsal spine is the summit of the back, but the third exceeds it in actual length, and measures 10·2 inches above the crown of the neural arch on its proximal edge and 9·6 on its distal one. Its greatest atlanto-sacral diameter is an inch and a half, that of the fourth and fifth neural spines being as much, while the posterior ones become narrower. The measurement of the spine was made exclusive of a thin peripheral epiphysis which has dropped off in the process of maceration from almost all the neural spines of the skeleton. The *sternum* in the musk-bull is divided into seven pieces or hæmal spines, the last or ensiform one being mostly cartilaginous. The first section or manubrium is capped on its proximal end by a large and not completely ossified piece, which coalesces with it, and furnishes a lateral pit on each side, to which a short broad hæmapophysis or sternal rib is articulated. The wide end of the first dorsal rib is connected partly to one of these sternal ribs, partly to the atlantal end of the terminal apophysis of the manubrium. Pits in the distal end of the sixth sternal piece receive the seventh pair of hæmapophyses; while the eighth pair, being adherent to them, are attached to pits common to the ensiform piece and the sixth. In the common ox the seventh and eighth pairs of hæmapophyses are merely approximated to each other, and the former is connected more exclusively with the sixth piece of the sternum. At the age of the bull (four to five years) the hæmapophyses are ossified, though not solidly.

Lumbars.

The *lumbars* are six in number.   As those of the adult musk-cow were detached and more convenient for examination and drawing than the lumbars of the skeleton of the bull, which could not be dismembered without injury, they were selected as the subjects of Plate XIV.   Fig. 1 gives a dorsal view of the four posterior ones; fig. 2, a side view of the three posterior ones; and fig. 3, 4, 5, different views of the fourth in the series.   The whole six measure, along the crests of the neural spines, 13 inches; and a quarter of an inch more on the sternal surfaces of the centra, allowance being made for the thickness of the intervertebral substances*.   All the lumbars are rounded on the sternal and lateral aspects of their centra, without any pinching in or vestige of a hypapophysial ridge; but the articulating discs are broader than the shafts, so that the latter are concave longitudinally. The transverse diameters of the discs considerably exceed the sterno-dorsal ones; and the lateral diameters of the first three shafts are about equal to one another; the three posterior ones increase successively in breadth, the last being also flatter sternad than any of the others.

The neural spines are nearly rectangular, thickened in the crests by the coalescence of the epiphyses, which have separated in the skeleton of the bull, and differ little from one another in height.   The dia-pleurapophyses are simple, stand out laterally with a slight inclination atlantad, and a scarcely perceptible one sternad.   They lengthen in succession from the first to the fourth, have a less atlanto-sacral diameter than in the domestic ox, and the last two are rounder and more curved atlantad than the others, to accommodate themselves to motion between the ilia.   There is no point on any of them resembling an anapophysis, though there are small angular ledges on the sacral edge of the neurapophysis near the zygapophysial articulations.

The zygapophysial joints fit with a sigmoid flexure, which is not so fully developed as in the fossil *Bison crassicornis* (Nos. 123, 124, 125, Plate XV.).   It is shown partially in the zygapophysis of the last lumbar of the musk-cow, Plate XIV. fig. 1, 2.

	LUMBARS.		
	FIRST.	FOURTH.	SIXTH.
Length of centra, excluding articulations	1·9 in.	1·8 in.	1·7 in.
Width of the antarticulating surfaces of centra	1·8	1·8	2·2
Sterno-dorsal diameter of ditto	1·3	1 3	1·3
Transverse diameter of the vertebræ between the prozygapophyses and the diapophyses	1·7	2·1	3·0
Transverse distance from the tip of one dia-pleurapophysis to that of the other, including the anchylosed epiphyses	6·1	8·6	7·8
Distance from sternal surfaces of centra to tips of neural spines, including coalesced epiphyses	4·2	4·1	4·1
From the most lateral part of one metapophysial tubercle to that of the other, transversely	1·7	2·2	3·0
Distance between the outsides of the zygapophyses	1·1	1·4	2·2
Length from the proximal part of the prozygapophysis to the distal part of the zygapophyses	2·8	2·9	2·8

* In the musk-bull the lumbars are connected by the intervertebral substances, which have shrunk in drying, and shortened that part of the spine unduly.   They still occupy a space of 14 inches, and in a recent state would fill probably 15 or 15½ inches.

Sacrum.

The *sacrals* may be enumerated as six, enclosing five pairs of foramina on both the dorsal and sternal aspects, for the transmission of nerves. The four proximal pieces are coalescent in the bull, and the two distal ones are connected by suture; but in the adult cow five have coalesced, though even in this animal the limits of the centra are clearly indicated by transverse lines. The lateral outlines of the sacrum are concave, from the breadth of the extreme pieces and the narrowness of the intermediate ones, while in the common ox and aurochs the sides are almost straight caudad of the articulations with the ilium. The neural spines of the three proximal sacrals are coalescent, and their summits form a smooth thickened and rounded ridge, except at the proximal corner, which remains thin.

The surfaces of the centra which form the floor of the neural canal of the sacrum differ from those of the common ox, aurochs, or fossil No. 89 (*Bison crassicornis*), in being flat and smooth, without the mesial ridge, which shows more or less in these, along with deep grooves ending in the foramina.

	MUSK-BULL, 4–5 YEARS.	ADULT MUSK-COW.
Width of the proximal end of the first sacral	6·2 in.	6·1 in.
Width of articulating surface of centrum of ditto	2·6	2·5
Width of middle of third sacral	2·1	2·0
Width of fifth sacral	3·0	2·9
Length of chord of six sacrals, sternal aspect	9·5	8·5
Height of neural spine from the sternal surface of first or second sacral	3·3	3·2

Caudals.

There are *six caudals* in the adult musk-cow, which measure together about 6½ inches in length. In the bull the last one and part of the penultimate one have been lost; the remainder measure more than 5½ inches. Four of them are flat on the dorsal aspect, with the broad thin diapophyses standing out laterally, like the same processes in the lumbars, and low neurapophyses rising at their proximal ends, but not forming an arch. The last two caudals are irregularly prismatico-conical, and are the only joints of the caudal series which are like those of the domestic ox. The sixth caudal scarcely projects beyond the tuberosity of the ischium.

Pelvis.

The pelvis differs from that of the domestic ox in many particulars, of which the form of the expansion of the ilium, the crest of that bone, the angle which the epiphysis of the ischium forms with the crus of that bone and of the pubal, and the forms of the protuberances on the dorsal edge of the ischium, are those which most readily strike the eye. In the following table of dimensions the measurements of the pelvis of the four- to five-year-old bull, the adult musk-cow, and of an aged Alderney cow are given. The skeleton of the latter stands at the same height at the shoulder with that of the musk-bull. Its several epiphyses and those of the musk-cow have coalesced with the bones to which they belong, but in the musk-bull the epiphysis which forms the crest of the ilium has dropped off in the process of maceration, while that of the is hium is partially coalescent.

M

	MUSK-BULL.	MUSK-COW.	ALD. COW.
Distance from the sternal angle of the crest of one ilium to that of the other	*13·0 in.	13·5 in.	18·3 in.
Transverse distance from the same point to the dorsal angle of the same ilium (or length of the sterno-dorsal chord of the crest)	7·4	7·7	9·1
Sterno-dorsal diameter of the iliac shaft at its narrowest place .	1·5	1·6	2·0
Sterno-dorsal diameter of the acetabulum	1·9	1·9	2·1
Atlanto-sacral ditto	2·0	2·0	2·0
Transverse distance from the apex of one lateral conical process near the dorsal angle of the ischium to the apex of the other	8·1	7·8	11·0
Transverse distance between the dorsal or spinous angles of the ischium, being the width of the pelvis there . .	4·7	5·0	8·2
Transverse diameter of the pelvis at the stem or ramus of the ilium	5·4	5·6	7·1
Length of foramen ovale	3·6	3·3	3·6
Sterno-dorsal diameter of ditto	2·6	2·1	2·3
Length of pelvis from the most proximal angle of the crest of the ilium to the tuberosity of the ischium	18·2	17·2	19·2
Distance from pubal brim of pelvis to tuberosity of ischium .	7·7	7·0	9·0
Distance from the dorsal angle of the ischium to the sternal symphysis of that bone	4·4	5·5	6·1
Distance from the dorsal angle of the ischium to the summit of its lateral conical process	2·0	1·7	3·2
Sterno-dorsal diameter of the dorsal ramus of the ischium .	1·4	1·7	2·6
Distance from the tuberosity of the ischium to the brim of the acetabulum nearest to it	7.4	6·8	8·5
Distance from the atlantal brim of the acetabulum to the sternal corner of the iliac crest	8·0	8·0	8·8
Length of the proximal ramus of the pubal from the edge of the acetabulum to the symphysis	2·8	3·1	3·5

The acetabulum has a more continuous brim than that of the domestic ox, the notch at the junction of the pubal with the ilium being filled up, while the one between the pubic and ischiatic portions of the brim remains as in the ox and most other *Bovidæ.* The ischiatic portion of the brim is likewise narrower and more defined, while in the domestic ox it is so much rounded off that it is not easy to ascertain the exact margin of the articular cavity after the bone has been macerated.

The sternal edge of the ilium, about an inch from the acetabulum, is less rounded than in the domestic ox, but not so acute as in the mustush or fossil bison. The pit mesiad of it also resembles that of the latter two species in depth, while in the ox it is a comparatively slight depression and wider. In the adult musk-cow there is a flat tubercle where the acute sternal edge of the ilium approaches the brim of the pelvis. This is just beginning to be developed in the four- to five-year-old musk-bull, but is not perceptible in the aged Alderney cow.

The crest of the ilium forms an arch with its convexity directed atlantad. The regularity of the arch is broken by a projection from it answering to the superior anterior spinous process in anthro-

* Without the epiphysis.

pology, and another marginal protuberance more iniad, which terminates the low rounded ridge that divides the lateral surface of the expansion of the ilium into two planes. In the domestic ox there is a wide shallow notch in the crest separating the dorsal and sternal angles of the bone into distinct lobes, the sternal one being much thickened, and including both the projections of the crest above alluded to. The dorsal lobe furnishes the articulating surface applied to the sacrum.

The *pubal* has a similar twist to that which it possesses in the domestic ox. In the adult musk-cow the epiphysis which forms the tuberosity of the ischium and the whole distal edge of the os innominatum is anchylosed to the ischium without a change of plane, but in the aged Alderney cow this process forms an angle with the symphysis of the pubis inclining dorsad. The dorsal angle and the lateral conical protuberance, both formed chiefly by this epiphysis, are much nearer to each other in the musk-cow than in the domestic ox.

Superior Extremities.

The *scapula* is flatter at the neck in the musk-ox and wider in the blade than that of the domestic cow. The spinal ridge is not quite so near the edge of the bone, and its distal or humeral end rises at a right angle without overhanging. The brim of the glenoid cavity is an ellipse, rather acute at the ends, but with a deficiency or wide notch behind the coracoid tubercle, which does not exist in the domestic cow.

	MUSK-BULL, 4–5 YEARS.	ADULT MUSK-COW.	AGED ALD. COW.
Length from the radial end of the glenoid cavity to the dorsal or shortest edge of the bone	13·6 in.	13·6 in.	14·3 in.
Length of the dorsal edge of the bone	8·3	8·1	7·4
Distance from the summit of the coracoid tubercle to the opposite or anconal margin of the glenoid cavity . . .	3·2	3·0	2·9
Width of the narrowest part of the neck of the bone . . .	2·1	2·1	2·2
Height of the front of the spine	2·1	1·8	1·3
Long axis of the brim of the glenoid cavity	2·4	2·3	2·2
Short diameter at its middle	1·8	1·6	1·9

The *humerus* greatly resembles that of the domestic ox, the slight differences in form of the tubercles not being easily described. The condyles of the musk-cow are more abbreviated, so that even the inner one falls short of the distal convexity of the articular surface, while it passes it some way in the Alderney cow. The length of the limbs of the musk-ox, in comparison with the bulk of its carcase, is greater than in the domestic cow.

	MUSK-BULL, 4–5 YEARS.	ADULT MUSK-COW.	AGED ALD. COW.
Length from the proximal edge of the greater tubercle to the most distal part of the elbow-joint	13·8 in.	12·7 in.	11·5 in.
Length from same point to the end of the inner condyle .	13·8	12·8	12·0
Length from the crown of the articular ball to the distal articular convexity	12·3	11·4	10·0
Circumference of the head and tubercles . . .	12·2	12·2	12·2
Circumference of smallest part of shaft . . .	5·1	5·1	5·2
Transverse diameter of the distal articulation . . .	2·8	2·8	2·9
Ancono-thenal diameter of inner condyle . . .	2·8	2·7	3·1
Ancono-thenal diameter of head and greater tubercle .	4·4	4·4	4·4

The *radius* of the adult musk-cow is as long as that of the Alderney cow, but is narrower and more arched on its rotular aspect.   The *ulna* however is shorter, though otherwise very similar.

	MUSK-BULL, 4–5 YEARS.	ADULT MUSK-COW.	AGED ALD. COW.
Length of ulna from the apex of the olecranon to the distal extremity that articulates with the cuneiform . . .	15·5 in.	13·9 in.	14·5 in.
Length of olecranon from the horn of the semilunar trochlea to the apex of the process	3·8	3·6	3·9
Popliteo-rotular diameter of the olecranon at the line of coalescence of the epiphysis	2·0	1·8	2·3
Its transverse diameter or thickness at the same place . .	1·0	0·9	1·0
Length of radius on its rotular aspect, from the most proximal brim of the elbow-joint to the distal articular surface of inner condyle	12·7	11·6	11·6
Transverse diameter of the elbow-joint	2·8	2·6	2·9
Transverse diameter at the tubercle below the joint . . .	2·9	2·9	3·1
Transverse diameter of carpal joint	2·5	2·3	2·4
Diameter at the coalescence of the epiphysis above the joint, including the ulna	3·0	2·8	2·7
Smallest breadth of rotular surface of the radius . . .	1·5	1·4	1·6

Differences between the radius of the musk-ox and the domestic ox may be readily detected in the elbow-joint.   In the domestic ox the radial part of the joint is divided into two lateral halves by an obtuse ridge, the mesial half being a simple cup, and the outer or radial one composed of a deeper and shallower trochlea.   In the musk-cow the dividing ridge is situated more mesiad or inward, is less elevated,—the cup on one side, and the trochlea on the other, being shallower.   The external trochlea is also narrower in proportion.   The prominence of the brim of the joint on its rotular aspect corresponds with the dividing ridge in both species.   There is also a well-marked distinction between the species in the articulation of the ulna with the radius.   Following the line of junction of the two bones as seen on looking into the elbow-joint, we may divide it in the domestic ox into three distinct parts.   The first part of the line, beginning on the outer side, runs transversely, with a slight convexity towards the ulna.   The next portion, diverging towards the olecranon at more than a right angle, is shorter.   The angle which the two lines would form at the point of meeting is truncated by a deep pit for lodging a lubricating gland.   The third part of the line of junction, about equal to the first in extent, is transverse, with a scarcely perceptible concavity towards the ulna.   Plate XV. fig. 5 represents this joint in the musk-cow.   The outer and second lines of junction both run obliquely rotulad, and form by their meeting an angular point, which coincides with the bottom of the greater trochlea for receiving the outer condyle of the humerus.   This point is not cut short by a pit, as in the domestic ox, but there is a dull spot on the articular surface of the trochlea pointing out the seat of a gland.   The inner portion of the joint is transverse and proportionally longer than the corresponding part in the domestic cow.   A deep pit in the olecranon for lodging glands and fat borders more than half the length of this line.   There is a pit also in the domestic ox, but it does not touch the edge of the radius at all.   The suture, which a projecting ledge of the ulna forms with a tubercle of the radius immediately below the outer margin of the elbow-joint, is carried out in the domestic cow as far as the outer border of the joint, or even beyond it in the large Lincolnshire oxen; but in the musk-cow it is almost half an inch further in. The fossil No. 72 (pp. 56, 57), figured on Plate XV. fig. 3, more resembles the common ox in this joint, but with some differences already pointed out.

The articulating surfaces of the radius and ulna, which contribute to form the carpal joint, are modelled much like those of the domestic cow.

The *carpal* bones are also very similar in the two species. The pisiform however is wider in the musk-cow, and the scaphoïdes and magnum also measure a little more transversely. The lunare like-wise presents some difference of shape on its popliteal aspect. The height of the scaphoïdes and mag-num, or of the lunare and unciform, is 1·6 in the musk-cow. The cuneiform and unciform measure a line more.

The *metacarpal* of the musk-ox is shorter and wider than that of the domestic cow, and also more flat and straighter on its popliteal aspect. The line indicating the confluence of the third and fourth metacarpals in the musk-cow is more indistinct than in the domestic species. In the aurochs there is a rudimentary fifth metacarpal, which reaches about one-third of the length of the cannon-bone, and a shorter ossicle representing the second metacarpal. In the common ox there is an oblong fifth metacarpal not above a fourth or fifth of the entire length of the cannon-bone, articulated to a small facet a short way below the outer corner of the surface on which the unciform moves. I have found no second metacarpal in the skeletons of the ox which I have examined, unless a small rising of bone completely coalescent with the inner proximal corner of the cannon-bone be considered as such; neither have I seen anything that can be considered as a second metacarpal in the musk-ox; but in the adult musk-cow there is a slender fifth metacarpal reaching past the middle of the cannon-bone, and articulated to a facet at its upper end, as in the domestic ox.

	MUSK-OX.	MUSK-COW.	ALD. COW.
Length from the ridge between the articulating discs at the proximal end of the cannon-bone to either of the ridges of its distal articulation	7·6 in.	6·9 in.	8·1 in.
Width of the proximal end of the cannon-bone	2·3	2·1	2·1
Width of the middle of its shaft	1·5	1·4	1·2
Width of the distal end of the bone	2·8	2·5	2·3
Length of fifth metacarpal	—	3·0	1·0

The *phalanges* of the two middle toes, which are all that are developed in the ruminants, are stouter in the musk-cow than in the Alderney cow; the first one is also longer, and the ungueal broader and less pointed. The four sesamoid bones at the metacarpal joints and the two at the ungueal ones are likewise larger in the musk-cow.

		MUSK-COW.	ALD. COW.
Length of the first phalanx		2·6 in.	2·2 in.
Ditto	second phalanx	1·5	1·5
Ditto	ungueal phalanx	2·4	2·4

Lower Extremities.

In the musk-ox the shaft of the *femur* is decidedly arched towards the rotular aspect, like that of the sheep, while that of the domestic cow is almost straight. The depression above the outer condyle does not extend so far, and it is limited above by a protuberance in the musk-cow which does not exist in the femur of the common ox. The edges of the rotular trochlea also are more nearly of the same height in the musk-ox, rendering the groove less oblique in its transverse section, and there is less tendency of the shaft to rise into a ridge where it joins the inner margin of the trochlea, as it does in some degree in the domestic ox, and gives a somewhat triangular section to that part of the shaft. In the musk-ox the section there is almost round. The other differences are slight, and can be better appreciated by comparison than description.

	MUSK-BULL, 4–5 YEARS.	ADULT MUSK-COW.	AGED ALD. COW.
Length from the crown of the ball of the femur to the distal curve of the inner condyle	14·4 in.	13·3 in.	13·8 in.
Length from the tip of the trochanter to the distal convexity of the outer condyle	14·8	13·6	15·0
Length from the crown of the ball to distal curve of the patellar trochlea	14·7	14·8	13·1
Width of the ball of the femur and shoulder of the trochanter	4·2	4·0	4·5
Width at the thickest part of the condyles	3·5	3·4	3·7
Rotulo-popliteal diameter of the inner condyle and the trochlea	3·5	3·7	4·1
Circumference of the shaft in its middle	4·8	4·5	5·0

As compared with the same bones in the domestic cow the *tibia* of the musk-ox is proportionally shorter than its femur. The ridge of the shin descending from the tubercle to which the ligament of the patella is attached wholly disappears in the middle of the shaft; but a raised line proceeding from the fibular angle of the knee-joint gradually curves forward on the rotular aspect of the bone, and ends at the brim of the trochlea, which moves on the outer arch of the astragalus. On the opposite side a shallow groove, bounded by two faint ridges or lines, proceeds from the popliteal hollow, and curving gradually outwards terminates in a notch of the distal articular surface behind the process of the inner malleolus. This latter groove is evident also in the domestic cow. The exterior distal surfaces which articulate with the fabella or distal rudiment of the fibula and the heel-bone differ in form from those of the domestic ox, and are not divided by a notch so completely into two distinct trochleæ.

	MUSK-BULL, 4–5 YEARS.	ADULT MUSK-COW.	AGED ALD. COW.
Length from crests in centre of the knee-joint for the attachment of the crucial ligaments to the most distal projection at the inner ankle	14·3 in.	13·2 in.	14·3 in.
Width of the head of the tibia	3·5	3·2	3·7
Width of its distal end	2·5	2·3	2·5
Length of the heel-bone	4·8	4·5	5·5
Greatest length of astragalus	2·25	2·2	2·75
Length of hind cannon-bone	8·3	7·1	9·3
Width of the middle of the shaft of ditto	2·3	2·3	2·1
Width of the distal end of ditto	4·8	4·5	5·2

The *metatarsus* or *hinder cannon-bone* is shorter and rather stouter than that of the domestic cow, and it has a less deep and distinct groove on its rotular surface, particularly towards its distal end, where it is comparatively flat. Its popliteal surface is nearly as straight and flat as in the domestic ox, showing little of the convexity in the middle of the shaft observable in the fossil metatarsus of *Bison crassicornis* or in the recent aurochs.

The *toes* of the hind foot are a little more slender than those of the fore one, but scarcely differ from them in length.

The skulls of the old musk-bulls show that an augmentation of osseous matter and consequent increase of the weight of the cranium goes on as age advances, but to what extent the old bulls surpass the cows of the herd in size we are unable, from want of data, to say. The preceding tables show that the musk-bull of four or five years of age has the general stature and length of limbs of an Alderney cow, which is one of the smaller of our domestic breeds. The musk-cow is less, and does not commonly surpass the stature of a Scottish highland kyloe. Lieutenant M'Clintock gives the following dimensions of the carcases of four musk-oxen shot by him on Melville Island.

	MUSK-BULL.	MUSK-COW.	MUSK-COW.	MUSK-COW.
From the horns to the root of the tail . . .	. 86·0 in.	70·5 in.	64·0 in.	62·0 in.
From the fore hoof to the top of the shoulder .	. 57·0	55·0	—	49·5
From the hind hoof to the top of the rump .	. 51·0	—	—	—
Length of tail	. 2·0	—	—	—
Length of one horn	. 27·0	24·0	—	19·0
From the top of one horn to that of the other .	. 32·0	27·3	—	27·5

The weight of the bulls killed by Sir Edward Parry's people on Melville Island exceeded 700 lbs., yielding about 400 lbs. of meat. The skin and head weighed 130 lbs. They stood 10½ hands high at the withers, or about 42 inches.

OVIS MONTANA, Geoffroy. *The Big-horn.* (Plate I.)

Description of the skeleton of a ram under two years of age, preserved in the Museum of Haslar Hospital. The animal was killed on the Rocky Mountains west of the Mackenzie, between the 66th and 67th parallels of latitude.

Skull.

Teeth : Incis. $\frac{0-0}{3-3}$; Canines, $\frac{0-0}{1-1}$; Perm. prem. $\frac{3-3}{3-3}$; True molars, $\frac{3-3}{3-3} = \frac{6-6}{10-10}$.

When this individual was killed its deciduous premolars had not wholly fallen out, the crown of the third mandibular one still remaining attached to its successor, and fragments also of the first and second maxillary ones still sticking in their sockets alongside the new teeth. Judging therefore from the order of dentition in a domestic sheep, we may fix the age of this big-horn ram at about twenty-two months.

The three *maxillary premolars* of the second or permanent series differ little from the homologous teeth of the common sheep, except in the lateral vertical folds or pillars of enamel so prominently and acutely developed in the latter, being less sharply defined in the big-horn. The first *true molar*, or fourth of the permanent series, wants the punctiform cemental pits (or occasionally pair of pits) enclosed by enamel, which are situated in the domestic sheep between the lobules of the crown, towards the mesial side; and instead thereof, the inflected horn of the anterior interior crescent of the tooth opens mesiad into the vertical sinus between the lobules. The other two true molars differ little from the corresponding teeth of the common sheep. In our big-horn, and in domestic sheep supposed to be of the same age, the third lobule of the last molar is just beginning to rise above its bony socket.

Mandible.—The *incisors* have been chipped on their crowns, so as to destroy the completeness of their form; but their shafts resemble those of the domestic sheep.   As in the latter animal, the third incisor and the incisoriform canine are much more slender than the two front pairs, and stand more posteriorly.   An anterior marginal vertical fold of enamel, which acute and prominent on the third or last *mandibular true molar* of the domestic sheep is indicated more slightly on the second molar, and still more faintly on the first, is considerably more indistinct in the several homologous teeth of the big-horn.   Differences are also perceptible in the crowns of the first and second molars of the two species; but at the early age of this young ram, the last molar is not worn down quite to the apex of the pillar in question.

The *mandible* of the big-horn is longer and stronger than that of the domestic sheep; the articular branch joins the middle part at an obtuse angle, and the anterior portion ascends more sharply.

Different breeds of the domestic sheep vary in their facial line.   In the hornless variety the arched form of the nasals causes the part between the fore borders of the orbits to appear sunk, beyond which the frontal rises and forms a rounded transverse ridge at the posterior third of the orbits, which is the most prominent part of the facial line.   From this the profile descends along the sagittal suture in an arch to the occipital crest.   Immediately behind the gibbous orbital plates the frontal is depressed and deeply pitted on each side of the mesial arch.   The parietals are convex transversely as well as longitudinally.   In the Iceland four-horned variety the frontal rises from the coronal aspect of the orbits as far forward as the supra-orbitar foramina, forming a transverse ridge, from which the mesial pair of horn-cores spring and take a more or less vertical direction, while a comparatively restricted surface on the inial bases of the orbital plates gives origin to the lateral or accessorial horn-core.   In the big-horn, the nasals being almost straight, there is no depression in the forehead, but there is a gentle rise of the frontal along the sagittal suture until it has passed the hinder borders of the orbits about an inch; the frontal then curves suddenly to attain the plane of the parietals lying at a right angle with the facial line.   The bone also swells up to unite with the base of the horn-cores from the supra-orbitar foramina,—on one side to the coronal suture, on the other and mesiad to within half an inch of the sagittal suture, in a manner which will be understood by an inspection of the several views of the cranium in Plate I.   The positions of the post-orbital depressions in the hornless sheep are entirely occupied by the horn-cores in the big-horn, which take in more of the frontal than the double pairs of horns of the Iceland sheep.   Much of the orbital plates is concealed by the lateral expansion of the cores, so that they appear to have little prominence; but in reality they have more, in regard to the encephalon, than in the hornless sheep, and their walls are convex and very thick.   In the young big-horn ram the frontal swelling from which the horn-core springs occupies more than one-third of the circumference of the orbit, and would doubtless extend further in an older animal. With a general direction of the horn-cores iniad, there is a lateral divergence, which in a length of seven inches places their tips nine inches apart, their bases being within less than two inches of each other; and from a basal circumference of nine inches they taper to less than three at the ends, which are blunt.   They are moreover three-sided, the sides being convex and the angles rounded off, without any sharp lines.   The coronal side is arched convexly in respect of its length, and the obtuse meeting of the two sides on the inial aspect concavely.   Of the three sides the mesial one is the flattest, and the coronal one most convex transversely.   The horn-cores of the Cashmere goat have the same relative origin and a somewhat similar direction, but a very different form, being in section only two-sided —viz., convex laterally, and flat mesially, with two sharp edges and a spiral acute point.   The horns of *Ovis Ammonoides* of Hodgson, preserved in the British Museum, have much resemblance in form to the American species; but both skull and horns of the Asiatic animal are greatly superior in size.

Like their cores, the *horn-sheaths* are three-sided, being compressed, flatly convex in front, with the broader and flatter sides meeting each other behind in a rounded edge. About twenty obtuse prominent wrinkles cross the front transversely, their ends swelling out beyond the planes of the sides. These wrinkles are less obvious and more distant and irregular towards the tip of the sheath. Very fine streaks or fibres mark the sides lengthwise, with only a few narrow, irregular, transverse folds. About five inches from the base of each horn-case there is a circular groove, like the joint of a reed, and another seven inches higher up, or about two inches from the tip. The tip itself is thin, compressed, but not pointed, and is concave on its mesial side. Scarcely half a circle is traced by the curve of the horn iniad and basilad, the radius of the curve on the convex side being 6½ inches, and of the opposite concave edge only 3½ inches. A spiral twist turns the horn at its point outwards laterally to more than the breadth of its base. The pair of sheaths weighs 41 oz. avoirdupois; but another pair, belonging to an older ram killed in the same district of the Rocky Mountains, weighs 142 oz.; and I have seen much larger ones than even these. These latter horns have each six rings like bamboo-joints, situated more remotely as they approach the tip, and perhaps marking the years of growth. In other respects the larger horns are like those of the younger animal, except that the curve extends to more than a complete circle, the tip in its spiral course passing about three inches beyond the base.

The sheath, along its convex curve, measures	34·5 inches.
Ditto, the concave curve	18·0
Its breadth at the base	3·3
Its circumference there	12·0
Circumference of the horns of the younger animal	11·0

The *parietals*, which in the domestic sheep are more or less convex longitudinally, and more boldly arched transversely, with an oblique general direction intermediate between iniad and basilad, are perfectly flat in their mesial portions in the big-horn; and their plane, which is the same with that of the adjoining inial part of the frontals, lies at right angles to the facial line. A raised line marks the boundary between the occipital plane of the parietal and the lateral part, which curves round above the squamosal to assist in forming the temporal fossa. The sagittal suture is obliterated between the parietals, and no line remains to point out the form and size of the interparietals; but the occipital suture can still be traced, though it seems to disappear more early than in the domestic sheep, in which it has a similar transverse direction. A shallow, wide temporal fossa, bounded above by the orbit and base of the horn-core, and on the basilar side by the zygoma and the acute edge of the squamosals, resembles that of the domestic sheep, and is dissimilar to that of the bisons and oxen. In the musk-ox the fossa has an intermediate form, approaching however considerably to the ovine type.

Two *superoccipital* surfaces meet in the very rough occipital ridge, the upper plane being flat and coincident with that of the parietals, and the other, which is convex transversely, having an oblique inclination antiniad and basilad towards the occipital foramen. A rough patch on each lateral corner of the upper plane, for the attachment of muscles, is better defined than in the domestic sheep. In oxen, bisons, and musk-oxen, the homologous rough places are below the edge of the occipital ridge, the change of situation following the difference of form of the super-occipitals, and the relations that their planes bear to the direction of the muscles which suspend or raise the head. There is no prominent occipital spine in the big-horn distinct from the ridge, though a small rough place marks its usual site.

In the *exoccipitals* to which the condyles and styloid processes belong, there is no material difference from those of the domestic sheep.   They have coalesced with the super- and basi-occipitals to the complete obliteration of the sutures, but the *paroccipitals* are still defined by suture on their exoccipital sides, though they have partially coalesced with the squamosals.   From the greater relative size of the condyles in the sheep, as compared with those of oxen, they are not so completely received into the articular cup of the atlas, and that joint therefore has more lateral motion.

The *basi-occipital* is similar in outline to that of the domestic animal, being about as broad as it is long; but it is flatter, and the lateral protuberances, which render its surface concave in the common sheep, are represented in the big-horn by an inconspicuous rough spot, while on the other hand the shoulders of the bone adjoining the condyles are well developed in the big-horn in accordance with the deeper notches of the atlas in which they move.   An acute hypapophysial ridge exists in the *basisphenoid* of the domestic sheep, but there is scarcely any perceptible ridge in the big-horn.

There is a difference between the species in the form of their *mastoids* and the inclination of the tympanic tubes.   No marked character offers in the *bony palate* distinguishing the species, but the premaxillaries are stronger in the big-horn, broader also in their middles, and united by a more curved suture to the maxillaries.

Some of the less highly bred varieties of the domestic sheep have nearly straight *nasals*, but these bones may be stated to be wider and shorter as well as less arched longitudinally in the big-horn.   A less deep depression before the orbit is formed by the lachrymal and malar in the big-horn, and the acute ridge of the malar does not approach so close to the margin of the orbits, as in the sheep.

The dimensions of the skull, measured by aid of callipers, are as follow :—

From the apex of the premaxillaries to the occipital plane	10·1 inches.
Width of the forehead between the orbits and horn-cores . . . · .	5·2
Width of the face at the maxillary eminences over the fangs of the first true molars .	3·3
Length from the inial end of nasals to the occipital plane	4·0
Height from the basilar edge of the occipital foramen to the facial plane between the horns	5·4

Spine.

The chord of the *cervicals*, measured over their neural spines, is 9½ inches, and the curve of the sternal surfaces of the centra, making some allowance for the shrinking of the intervertebral substances, nearly 13 inches.   In a domestic sheep the latter measurement was less than 12 inches. On comparing the *atlas* (Plate XII. fig. 5, 6) with that of the common sheep, the similarity is obvious, but the table of dimensions will show some differences in the relative proportions of the several parts; and the greater depth and regularity of the notches for receiving the shoulders of the basi-occipital, as well as of the lateral notches between the prozygapophysial articular surfaces, and those of the centrum, readily strike the eye.   These lateral notches do not, as in the case of the musk-ox, furnish fulcra for prolongations of the condyles, for their edges are acute, but they admit of more lateral motion than in the domestic sheep, where the notches are very imperfect.   The process that separates the lateral from the sternal notches is narrower and more prominent in the big-horn than in the sheep, also more rounded on the summit and equally strong; it is in fact nearly semi-orbicular, but has a rather acute margin, and is concave within.   Between the notches on the sternal border of the antarticular cup there is a low ridge, divided at its summit by a very narrow furrow. This is shown in fig. 5, which gives an oblique view of the sternal aspect. The hypapophysial spine does not pass the postarticular surface as it does in the sheep, and the centrum of the big-horn is relatively longer, the projections sacrad of the ends of the lateral processes being shorter. On the other hand the neural spine is less prominent in the big-horn than in the sheep, but the surface of the centrum that faces the neural canal is rough as in the musk-ox, though not to the same degree. Thin epiphysial borders of the lateral processes have dropped off in maceration, so that the figures do not present the exact outline of the mature bone.

	BIG-HORN.	SHEEP.
Transverse diameter of the antarticulating cup of the atlas at the prozygapophyses	2·4 in.	2·1 in.
Sterno-dorsal ditto on the mesial plane, including the neural arch	1·3	1·1
Width at the shoulders of the lateral processes	3·1	3·5
Ditto near the sacral ends	3·0	3·2
Transverse diameter of the postarticulation	2·3	2·0
Distance between the arterial foramina, sternal side	1·7	1·9
Length of centrum on the mesial line	1·4	1·1

As compared with its homologue in the sheep, the *dentata* of the big-horn has a rather taller odontoid process, a wider antarticulating surface, a more rounded and fuller centrum, and less projecting diapophyses. In the sheep the mesial line of the centrum is acute in the middle, and it is terminated sacrad by a decided knob; while in the big-horn there is merely a more wide rounding of that end of the bone. The neural spine is thinner than in the sheep, and does not swell out to such a degree towards the distal end of its crest.

	BIG-HORN.	SHEEP.
Length from the edge of odontoid to the distal point of the centrum, mesial plane	2·8 in.	2·6 in.
Length from same point to the tips of the parapophyses	3·0	3·0
Length from the same point to the distal edges of the zygapophyses	3·3	2·8
Length of crest of neural spine	2·1	2·2
Sterno-dorsal diameter from front of the centrum to proximal end of neural spine	2·1	2·0
Ditto, ditto, distal end	2·5	2·7
Transverse diameter of proximal articular surface	2·3	2·0
Height of odontoid process	0·7	0·5
Length of centrum, excluding that process	2·1	2·1

The *fourth, fifth,* and *sixth* cervicals of the big-horn differ more from the homologous bones of the domestic sheep; their neural spines being more slender and tapering, but not longer, their centra longer, and in the fifth and sixth showing less of a hypapophysial ridge with its peaked distal termination than in the common sheep. The diapophyses and parapophyses are also thinner in the big-horn, their ends less dilated, and the bases of the third and fourth more connected by a thin plate of bone, the notch between them being slight compared to that in the sheep.

	THIRD CERVICAL.		FOURTH CERVICAL.	
	BIG-HORN.	SHEEP.	BIG-HORN.	SHEEP.
Length of centrum, sternal side, excluding the articulations	1·5 in.	1·3 in.	1·4 in.	1·3 in.
Length from the proximal edge of the prozygapophysis to the distal edge of the zygapophysis	2·3	2·1	2·1	1·9

	THIRD CERVICAL.		FOURTH CERVICAL.	
	BIG-HORN.	SHEEP.	BIG-HORN.	SHEEP.
Length from proximal tip of parapophysis to the distal end of the diapophysis	2·1 in.	1·9 in.	2·0 in.	1·8 in.
Width on the outsides of the arterial foramina, proximal end	1·3	1·3	1·2	1·3
Breadth between the outsides of the prozygapophyses .	1·7	1·8	2·0	1·9
Sterno-dorsal diameter from the sternal side of the centrum, at its proximal end, to the summit of the spine . .	2·1	2·1	1·9	2·2

In the domestic sheep the *sixth* and *seventh* cervicals have a faint hypapophysial ridge, and the ends of their neural spines, diapophyses, and pleurapophyses are, like those of the preceding vertebræ, considerably dilated. The ridge is imperceptible in the homologous cervicals of the big-horn, and the processes in all the cervicals are simpler.

	CERVICALS OF BIG-HORN.		
	FIFTH.	SIXTH.	SEVENTH.
Length of centrum, sternal side, excluding the articulations . .	1·5 in.	1·4 in.	1·0 in.
Length from proximal edge of the prozygapophysis to the distal edge of the zygapophysis	2·1	1·9	1·7
Length from the proximal tip of the parapophysis to the distal end of the diapophyses	1·6	1·2	—
Width on the outsides of the arterial foramina, proximal end	1·6	1·6	1·7
Breadth between the outsides of the prozygapophyses . . .	2·1	2·1	2·2
Sterno-dorsal diameter from the sternal side of the centrum at its proximal end to the summit of the spine	2·1	2·2	3·0

The length of the thirteen *dorsals*, measured from the tip of the first neural spine to that of the last, is 12·3 inches, and along the sternal aspect of their centra fifteen inches, no allowance being made in either case for the shrinking of the intervertebral substances in drying. The centra become successively narrower on their sternal aspects; the first has a broad, flat, sternal surface; in the third the sides are visibly pinched in, and a hypapophysial ridge is formed; in the eighth, ninth, and tenth, the compression of the sides attains its maximum, the posterior centra augmenting a little in breadth and length. All have the mesial line on the sternal aspect more or less acute. None of the cervicals are furnished with any metapophysial projection. That process is first developed in a tubercular form on the diapophysis of the second dorsal: it increases in prominence in the succeeding vertebræ, at the same time receding from the end of the diapophysis to which the tubercle of the rib is articulated. In the twelfth it projects from the under part of the prozygapophysis, and in the thirteenth it is a smooth rounded thickening of the back of that process. The twelfth and thirteenth ribs want the tubercles, and consequently the articulation with the diapophyses, and their prozygapophysial joints assume the lateral position of those of the lumbars, a change also taking place in the form and direction of their neural spines, which, instead of inclining sacrad like the other dorsal spines, are vertical, like the lumbar ones. The fourth neural spine stands highest in the skeleton; but the third is not inferior to it in actual length, while the thirteenth is almost thrice as short. In the domestic sheep the neural spines and diapophyses are stouter and thicker at the ends; and the metapophysis is much more prominent, and distinctly formed, having in the fourth and more posterior dorsals a length equal to that of the diapophysis itself. In the eighth and the succeeding ones it has approached the prozygapophysial joint so nearly as to furnish it with much lateral support, which it scarcely does in the big-horn.

	DORSALS.		
	FIRST.	SEVENTH.	THIRTEENTH.
Length of centrum, sternal side, excluding articulations . . .	0·9 in.	1·0 in.	1·3 in.
Width at the distal cup for the head of the rib, or behind the diapo- physes	1·4	1·2	1·1
Height of the neural spine above the neural arch, proximal edge .	3·6	3·8	1·5
Height of fourth and highest neural spine . . . 4·4 inches.			

The *ribs* of the big-horn are more slender and rounder at their dorsal ends than those of the domestic sheep, and also thinner and rather more expanded towards their sternal ends. The longest are the seventh, eighth, and ninth. The eighth and ninth have also the longest hæmapophyses or sternal ribs, all of which contain much granular ossific matter.

The *sternum* consists of seven bony pieces, with a broad membrano-cartilaginous appendix to the last one. A corner of the broad end of the first rib is articulated to an epiphysis on the dorsal end of the *manubrium sterni*, and the rest of that end of the rib is joined, through the intervention of a short broad hæmapophysis, to a lateral pit at the same end of the manubrium. The succeeding hæmapophyses articulate with cups common to two adjoining pieces up to the seventh and eighth, which have a common connection with the cup formed by the sixth and seventh pieces.

All the *six lumbars* have an acute hypapophysial line curved concavely, and are also concave on the sides. Their prozygapophyses have thick, convex, smooth backs, which assume the prominence of tubercular metapophyses in the penultimate and antepenultimate one. There is no metapophysis whatever on the sixth. The dia-pleurapophyses have a slight inclination atlantad and sternad, as well as laterally. The last one curves more regularly atlantad. In the number of dorsals and lumbars the Cashmere goat agrees with the big-horn, each having nineteen in all, which seeems to be the normal number among the *Bovidæ* and *Ovidæ*.

	LUMBARS.		
	FIRST.	FOURTH.	SIXTH.
Length of centrum, sternal side	1·3 in.	1·3 in.	1·1 in.
Length from the proximal edge of the prozygapophysis to the distal edge of the zygapophysis	1·9	2·2	2·0
Transverse distance from the back of one prozygapophysis to that of the other	1·3	1·6	1·9
Transverse diameter behind the diapophysis	1·1	1·2	1·5
Sterno-dorsal diameter from the sternal side of the centrum to the crest of the neural spine, proximal edge	2·8	2·7	2·6

The *sacrum* consists of five pieces, enclosing four pairs of holes, and in process of anchylosis. Their spines are coalescent at their bases, and the crests of the four anterior spines are ossifying from two centres in each, a right and a left one separated by a mesial line of cartilage.

Seven coccygeals or *caudals* a little exceed the five sacrals in length, and much resemble those of the musk-ox. The three last are subcylindrical, the others depressed, and the last one projects an inch and three-quarters beyond the tuberosity of the ischium.

Length of the chord of five sacrals, sternal aspect	4·5 inches.
Length from prozygapophysis to distal corner of fifth spine	5·1
Length of seven coccygeals	6·2

Fore Extremities.

The disproportion in the length of the limbs between the big-horn and domestic sheep is greater than that of their spinal columns. In outline and in the form and position of the spine the *scapula* closely resembles that of the musk-ox. The coracoid process has a thinner edge, more incurved on the mesial or radial aspect; its shape on the ulnar side is the same as in the musk-ox, but the glenoid cavity is less elliptical, being subrotund, except that an angle of the articular surface projects on the base of the acromion; the notch within that process on the mesial side of the brim is deeper and narrower than in the musk-ox. The homologous bone of the domestic sheep has a broader base in proportion to the length of the other sides, and its glenoid cavity is less rotund, being somewhat oblique in the anconal half of its margin. In other respects it agrees with the scapula of the big-horn.

	BIG-HORN.	SHEEP.
Length of thenal or coracoid edge of the scapula, excluding cartilage	7·9 in.	6·1 in.
Ditto of anconal edge	7·7	5·9
Length of dorsal edge or base	5·2	5·0
Height of acromion above the blade of the scapula	1·2	1·0
Breadth of cartilagino-osseous appendage to base, when recent	1·6	2·0

Scarcely any perceptible difference of form exists between the *humerus* of the big-horn and that of the sheep, except that the acute edge or ridge, which in the latter descends from the greater tubercle down about two-thirds of the length of the shaft, is smooth and rounded in the big-horn, but yet that side is narrower than the anconal one. It is probable that in an older animal the ridge may be more acute; in our specimen the tubercle has not completely coalesced with the shaft.

	BIG-HORN.	SHEEP.
Length from the crown of the ball to the extremity of the inner condyle	7·6 in.	5·3 in.
Length from the tip of the greater tubercle to the extremity of the outer condyle	8·5	5·9
Diameter of the head and greater tubercle	2·5	1·9
Transverse diameter of the condyles	1·6	1·2

As compared with its homologue in the ox or bison the *radius* is narrow, thin in its ancono-thenal diameter, and arched thenad still more than in that of the musk-ox. Its *ulna* also resembles the last-named species in the form of the olecranon, but the distal half of the bone is depressed, applied very flatly to the radius, and somewhat removed from the edge of that bone.

In the course of the line of junction of the ulna and radius, which traverses the elbow-joint, the big-horn approaches to the musk-ox (Plate XV. fig. 5), but differs characteristically from it, and still more widely from the homologous part of the oxen or bisons. Beginning on the outer or lateral side we have first a bold curve convex towards the olecranon, then a narrow process of the ulna rounded at the tip, running thenad in the outer trochlea of the radius; followed by a narrow process of the radius entering the ulna, contiguous to a pit for lodging lubricating glands; lastly the line between the radial side of the ulna and the inner trochlea of the radius is very slightly concave towards the former. This joint in the domestic sheep differs from that of the big-horn in the lateral half, the lines bounding the lateral process of the ulna forming of it a nearly equilateral acute spherical triangle, whose apex does not dip so far thenad into the radius as the narrow but obtuse one in the big-horn. There is a difference also in the form of the point of the radius which follows, but the mesial half of the joint is alike in both species.

Length of the radius on its thenal aspect and mesial line . . . . . . 9·0 inches.

Width of ditto at its proximal end . . . . 1·7

Width of the middle of the shaft . . . . 1·0

Theno-anconal diameter of its middle . . . . 0·5

Length of the ulna from the tip of the olecranon to its distal end (its chord) . . 11·1

Length of olecranon from the horn of the crescentic curve of the elbow-joint . . 1·4

Scarcely any difference can be detected in the forms of the *carpals* from those of the musk-ox, except in the pisiform, which is somewhat lozenge-shaped in the big-horn, but more orbicular in the musk-ox. Both are convex exteriorly and concave on the mesial aspect. The height of the scaphoid and magnum conjointly is an inch.

In length the compound *metacarpal* or cannon-bone rather exceeds that of the Alderney cow, is but little shorter than that of a full-grown Spanish ox, and considerably exceeds its homologue in the musk-ox. It is however slender, has a much more convex thenal surface than in the *Bovidæ* just named, and is quite straight on the anconal side, like the metacarpal of a musk-ox, not gibbous in the middle, like that of the bison. Like most of them it is concave transversely at the proximal end and for nearly half-way down on the anconal side. Scarcely a trace of a mesial line of division exists between the coalesced third and fourth metacarpals. The mesial groove is also almost wholly obliterated in the cannon-bone of the adult musk-cow.

		BIG-HORN.	SHEEP.
Length of the metacarpal		7·5 in.	5·4 in.
Its width at the carpal joint		1·3	1·2
Width in the middle		0·75	0·78
Ditto at the distal end above the joint		1·5	1·3

A filiform rudiment of a fifth metacarpal is articulated to the cannon-bone nearly half an inch below the carpal joint. In the domestic sheep the rudimentary fifth metacarpal is longer than in the big-horn, and equally slender; and there is also a slender styliform second metacarpal, not so long as the fifth one, and, like it, attached below the joint. In a three- or four-year-old sheep the proximal end of this ossicle is coalescent with the third metacarpal; but in our big-horn no vestige of it remains, nor any facet to indicate its point of attachment. Instead of a convex expansion laterally, as in the oxen, the ungueals are much compressed, being wedge-shaped and as flat on the outer as on the inner side. The sesamoid bones belonging to the metacarpal joint form more perfect trochleæ for the tendons on their posterior surfaces than their homologues do in the *Bovidæ*. The grooves they form on their inner sides are also narrower and deeper, in accordance with the more acute articular ridges of the metacarpal.

Length of the first phalanges of the toes . . . . 2·2 inches.

Ditto second ditto . . . . 1·1

Ditto ungueal ditto . . . . 1·1

Ancono-thenal axis of the ungueal . . . . 1·7

Hind Extremities.

Though more slender than the *femur* of the musk-ox, that of the big-horn greatly resembles it in form, and in the amount of its convexity rotulad; it is considerably less arched than its homologue in the domestic sheep. The condyles, like those of the sheep, are more rounded than in the ox.

Length from the ball of the femur to the distal convexity of the inner condyle . . 10·1 inches.
Length from ditto to the distal curve of trochlea . . . . 9·9
Length from tip of the trochanter to convexity of outer condyle . . . . . 10·4
Diameter of the trochanter and articulating ball . . . . 2·2
Transverse diameter of the two condyles . . . . 2·1
Rotulo-popliteal diameter of the inner ridge of the trochlea and inner condyle . . 2·5
Rotulo-popliteal diameter of the outer ridge of the trochlea and the outer condyle . . 2·1
Circumference of the shaft in its middle . . . . 3·0
Its lateral diameter there . . . . 0·87
Rotulo-popliteal ditto . . . . 1·05

Except that it is much more slender and proportionally longer, the *tibia* has most of the characters of that of the musk-ox. It is however less concave and flatter on the popliteal aspect near the knee-joint, and rounder on the same side in its distal half. In both species the tibia is arched popli-tead, and rather more so in the big-horn.

Length of the tibia of the big-horn from the crests in the knee, to which the crucial
    ligaments are attached, to the tip of its outer malleolar process . . . . 12·7 inches.
Its width at the knee . . . . 2·3
Its width at the ankle . . . . 1·4
Length of the calcaneum . . . . 3·3
Length of metatarsal . . . . 8·2
Width of the metatarsal joint . . . . 1·4

In length of the *metatarsal* the big-horn absolutely excels the musk-ox and comparatively the ox or bison. It differs from its homologue in the musk-cow or common ox in the greater prominence of its inner popliteal tarsal angle, and in the consequent obliquity of the upper half of that surface of the bone; and the bone is as slender as the other parts of the limb.

Length of the first phalanx of the hind toes . . . . 2·2 inches.
Length of the second ditto . . . . 1·4
Length of the ungueal . . . . 1·1

Height of the prepared skeleton at the fourth dorsal spine, with the limbs resting
    in a natural position on the ungueals . . . . 35·7 inches.
Length from between the horns to the tip of the coccygeals; the neck raised, and no
    allowance made for the shrinking of the intervertebral substances, by which the
    vertebræ are still attached to each other, curve of the neck included . . . 51·0
Chord of the distance between the proximal end of first dorsal and the tip of the last
    coccygeal, with the intervertebral substances contracted . . . . . 33·8
Allowing for 3·4 inches' shrinking of cartilages, the same points would be distant from
    each other about . . . . 37·0

In the 'Fauna Boreali-Americana' (vol. i. p. 274) a male and a female big-horn killed on the Rocky Mountains near the 62nd parallel are described. The male was older than the one whose skeleton is described above, stood forty-one inches high at the shoulder before it was skinned, and each of its horns measured at the base thirteen inches in circumference. The old rams are said to be almost totally white in winter. I was not informed at what season the one whose skeleton forms the subject of the preceding pages was killed. Its coarse cervine hair is on the back dark brown, mixed with a few white hairs, which become more numerous as the distance from the dorsal line increases, rendering the flanks grizzled. The dark brown extends forward to the occiput. Between the ears the hairs are rather long, and stand up irregularly, the white ones greatly predominating there. On the face they are brownish-white, and short and grizzled on the sides of the mandible and adjoining parts of the neck, the brownish-black and white hairs being intermingled there. The dark brown extends to the fore-shoulder, with a sprinkling of white hairs, and the outsides of the fore and hind legs are mostly blackish-brown of a purer tint, but fading in various directions to brown. Pale wood-brown hair covers the sternum, and the belly and insides of the legs are pure snow-white. The white passes up the inner sides of the thighs and changes to a soiled or brownish-white on the buttocks, which have the circumscribed light-coloured patches observed in the rein-deer, wawaskeesh, or wapito, and many other cervine animals. The short tail is dark brown, and a brown mesial line running from it interposes between the whitish buttocks and loses itself in the dark tints of the back.

There is no naked skin about the mouth, except the very narrow margins of the nostrils. The hairs curl inwards over the borders of the lips and also across the mesial line between the mouth and septum of the nose, the cutis scarcely showing among them. The ears are covered densely within and without with hair which, contrasted with that of the body, is comparatively fine and flexible. Within the ears are white, on the outsides grizzled, the black hairs greatly predominating. The scrotum is prominent, not pendulous, and is thickly covered with white hairs. There is a space on each groin also which is thinly clothed with short appressed hairs, and is the most naked part of the whole hide.

Blasius is reported to have mentioned at the meeting of naturalists at Brunswick, the following generic distinctions between *Ovis* and *Capra*. All sheep possess a distinct lachrymal groove which is wanting in the goats. The forehead in *Capra* rises to a steep protuberance; in *Ovis* it is flat or even somewhat hollowed. In all species of *Ovis* the greatest diameter of the horn is across the longitudinal axis of the head, while in all species of *Capra* it runs parallel with it. In *Capra* the hoof, viewed sideways, is scarcely higher before than behind; in *Ovis* it is triangular, running to a point posteriorly like a goat's hoof cut diagonally. Specific distinctions, he also remarked, might generally be found in the arrangement and direction of the horns. In *Ovis argali*, *O. montana*, and *O. nahor*, the right horn winds in a space to the left, and the left horn to the right. *O. nivicola* and *O. Californica* are identical with *O. montana*. In *O. musimon*, and *O. Vignii* of Hodgson, the horn has the same twist, but so slight, that the anterior surface lies quite on the same level, and the twist is only perceptible on the posterior surface. In the *O. tragelaphus*, *O. orientalis*, *O. burhel*, and *O. cyprias*, the left horn is twisted to the left, and the right to the right: the direction of the tips and spread of the horns is affected by this twist. Our domestic sheep, in respect to the form of the horn, comes nearest to *O. musimon*, and to an undescribed species signalized by Brandt. (Report on Zoology in 1842, by Prof. Andr. Wagner; Ray Society, 1845.)

In applying these rules to the big-horn seriatim, the lachrymal bone exhibits nothing like the cervine sinus in either the big-horn, domestic sheep, or goat. A canal, commencing by two openings on the orbital edge of the lachrymal, runs within the substance of the bone into the cavity of the nose in the big-horn, as well as in the domestic sheep. The forehead of the big-horn rises steeply to its summit between the horns, in which it responds to the character of *Capra*, as does also the direction of the long axis of a section of the horn. This is still more evident in the compressed horn-core,

o

whose greatest diameter at the base, though not exactly parallel to the mesial plane of the skull, is but slightly inclined to it, and forms a very large angle with the transverse diameter of the skull. In the domestic sheep, on the contrary, the base of the horn has a decidedly transverse direction, though it does not meet the mesial plane quite at a right angle. The hoofs of our big-horn have been removed with the skin, but both fore and hind ungueals resemble those of the sheep in profile. As to the horn-cases, the right one of the big-horn certainly winds spirally to the right, and the left one to the left, without any change of the relative aspect of the front or regularly barred side of the case. There is however no twist whatever in the core, its direction being as stated in page 88. If therefore there has been no mistake in this part of Blasius's observations, he must mean something which I do not comprehend. In the position of the great diameter of the horn the big-horn has the caprine character; yet it is ranged as a sheep by Blasius himself.

The big-horn ranges northward to at least the 68th parallel of latitude on the Rocky Mountains, and southward on the same chain to California, if the identity of species of the northern and southern ones be correctly established by Blasius. It never descends into the low country east of the Rocky Mountains, but is, I believe, an inhabitant of some of the highlands which intersect the country bordering on the Pacific coast.

Table of Dimensions of two Rein-deer Skulls of the Barren Ground species from Great Bear Lake.

	ADULT MALE.	ADULT FEMALE ?
Length from the antinial tip of the premaxillary to the occipital spine, by callipers	15·1 in.	14·1 in.
Length from the antinial end of the nasals on the mesial plane to the occipital spine, by callipers	11·3	10·6
From inial end of nasals on the sagittal suture to the occipital spine	6·8	6·2
From the antinial tip of the maxillary to the antinial edge of the orbit or lachrymal	9·8	8·8
From the inial edge of the orbit to the occipital spine	5·2	5·1
Lateral diameter across the frontal from the rim of one orbit to that of the other on a line with the super-orbital foramina	5·2	4·9
Lateral diameter between the borders of the two orbits at the junction of the frontal and malar bones, being the widest part of the skull	6·4	6 2
Lateral diameter between the lateral edges of the squamosals, above the auditory canals	5·0	4·8
Distance from the lateral edge of one occipital condyle to that of the other at their bases	2·6	2·7
Height of occiput from basilar edge of the great foramen to the coronal edge of the occipital ridge	3·0	3·1
Height from basilar edge of the occipital foramen to the convexity of the frontal between the antlers and before the coronal suture, by callipers	4·0	3·9
Length of the chord of the premaxillary	4·4	4·0

	ADULT MALE.	ADULT FEMALE ?
Length of the nasals (mesial processes). The lateral antinial processes of these bones are slightly longer than the mesial ones	4·5 in.	4·4 in.
Length of mandible from the cutting edge of the incisors to the angle of the jaw	12·0	11·2
From incisors to posterior edge of mandibular condyle	12·8	11·8
From incisors to tip of coronoid process	13·3	12·4
Width of the face at the protuberances of the maxillaries above the fifth molars	4·2	4·1
Width at the maxillary canines	2·6	2·5
Length of the turbinal plate which interposes between the premaxillary and nasal	1·3	0·9
Greatest breadth of the crowns of the three true maxillary molars	0·45	0·44
Length of space occupied by the six maxillary molars	3·5	3·7
Distance from the first premolar to the permanent small canine implanted in the apex of the maxillary	2·5	2·4
Distance from the canine to the apex of the premaxillary	2·8	2·3

As the preceding sheets were passing through the press, the skeleton of the aurochs in the British Museum was taken down for the purpose of being cleaned, by which I have had an opportunity of examining some parts not previously accessible, and among others the radio-ulnar articulation as seen on looking into the elbow-joint. In the aurochs this joint much more closely resembles that of the domestic ox than it does that of *Bison crassicornis*, described in page 57, and figured on Plate XV. fig. 4. As compared with the latter, the exterior or lateral portion of the line of junction is a gentle curve, convex towards the olecranon, while in the fossil it is straight; the next line, or that which ascends obliquely towards the olecranon, is the same in both species, while the third portion is straight in the aurochs, but in the fossil is defined by the meeting of two legs of a very obtuse angle, opening towards the olecranon. There is also a difference in the lateral portion of the articular surface of the radius, being more convex transversely in the aurochs, and requiring a deeper trochlea in the outer condyle of the humerus to fit it than in the fossil, in which that part is flatter. The differences that exist between the radio-ulnar joint of the aurochs and of the domestic ox are comparatively slight. The mesial end of the lateral portion does not dip so much into the radius in the aurochs as it does in the ox, and there is a very much smaller depression there for lodging glands or fat; and the third, or inner section of the joint, is longer in proportion in the aurochs. The following dimensions of the radius and ulna of this adult bull-aurochs may be compared with those of the homologous bones of the *Bison crassicornis* and Lincolnshire ox as given in page 57 :—

Length of the radius, from the proximal anconal edge of the elbow-joint to the most distal point of the carpal joint, inner side	13·2 inches.
Transverse diameter of the radial part of the elbow-joint	3·4
Transverse diameter of the radius at its protuberance, immediately below the elbow-joint	3·9

Transverse diameter of the shaft of the radius near its middle . . . . . 2·3 inches.
Transverse diameter of the radio-ulnar surface articulating with the carpus . . 3·3
Ancono-thenal diameter of the radius near the middle of its shaft . . . . 1·4
Length of the ulna from crest of the olecranon to outer tip of the carpal joint . . 17·8
Ancono-thenal diameter of the olecranon where greatest, or at the crescentic process
of the ulna . . . . 3·5
Ancono-thenal diameter of the olecranon, on a line with the proximal brim of the
radius* . . . . 2·2

In the almost regularly oval glenoid cavity, the scapula of the aurochs differs from that referred to *Bison priscus* in page 39, wherein it is ovate.

* The measurement of the process at the same place is, in the fossil 2·8 inches, and in the domestic ox 2·0 inches.

Haslar Hospital, October 1, 1852.

ERRATA.

Page 45, No. 115, for *Third cervical*, read *Fourth cervical ;* and Page 47, No. 116, for *fourth cervical*, read *fifth cervical*. Also in Page 49, line 9, *for* fifth, *read* third.

These corrections imply also some reformation in the comparisons with the homologous parts of the domestic ox. The dimensions of the fourth and fifth cervicals of the latter animal are to be found in page 48.

Plate I.

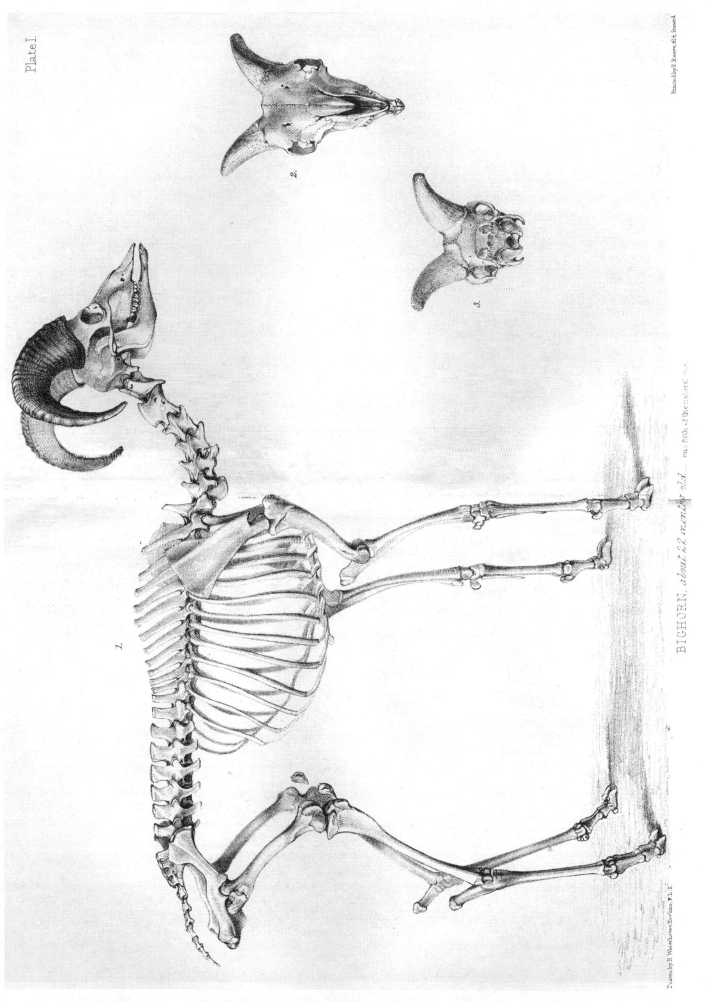

BIGHORN. *about 2.2 months old*.... *one-fifth of the natural size*.

Printed by F. Reeve, 64, Strand.

Drawn by B. Waterhouse Hawkins, Pl. I.

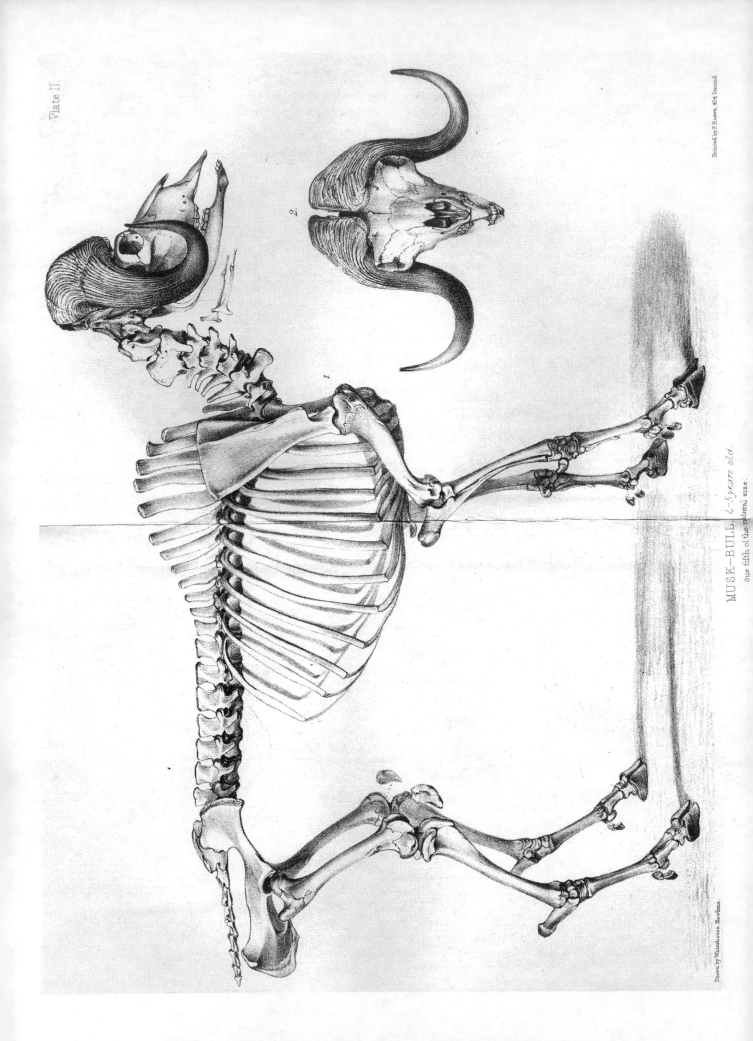

Plate II.

MUSK–BULL. *4.–5 years old.*
one fifth of the natural size.

Drawn by Waterhouse Hawkins.

Printed by P. Reeve, 314 Strand.

MUSK–BULL. 4–5 years old.
Nuchal aspect of the Skull.
Natural size.

The material originally positioned here is too large for reproduction in this
reissue. A PDF can be downloaded from the web address given on page iv

Plate IV.

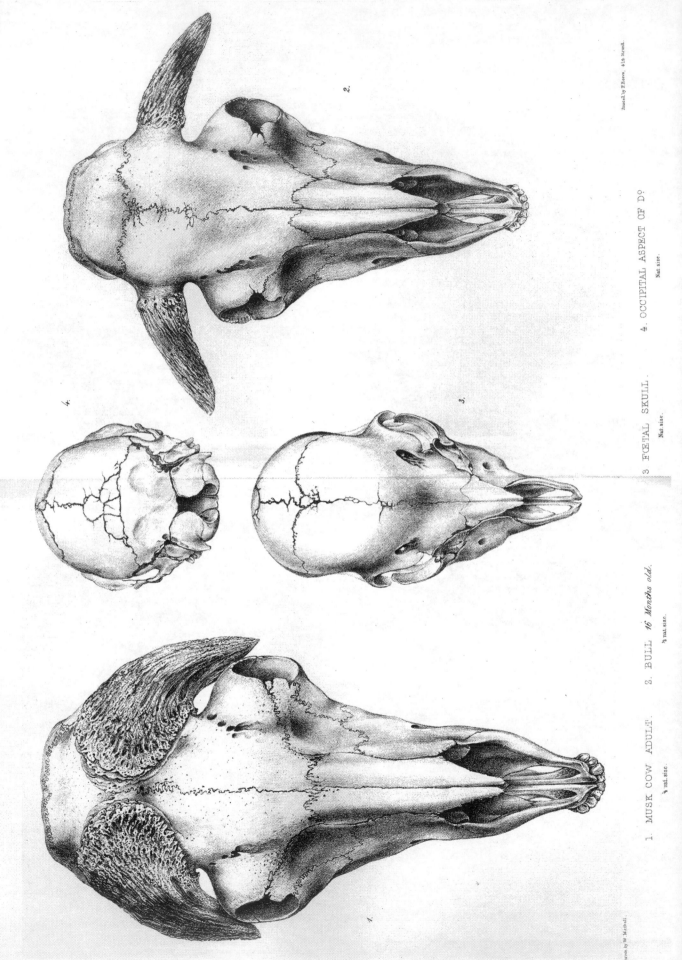

1. MUSK COW ADULT. ⅓ nat. size.   2. BULL *16 Months old.* ⅓ nat. size.   3. FŒTAL SKULL. Nat. size.   4. OCCIPITAL ASPECT OF D°. Nat. size.

Drawn by W. Mitchell.

Printed by J Reeve. 4½ Strand.

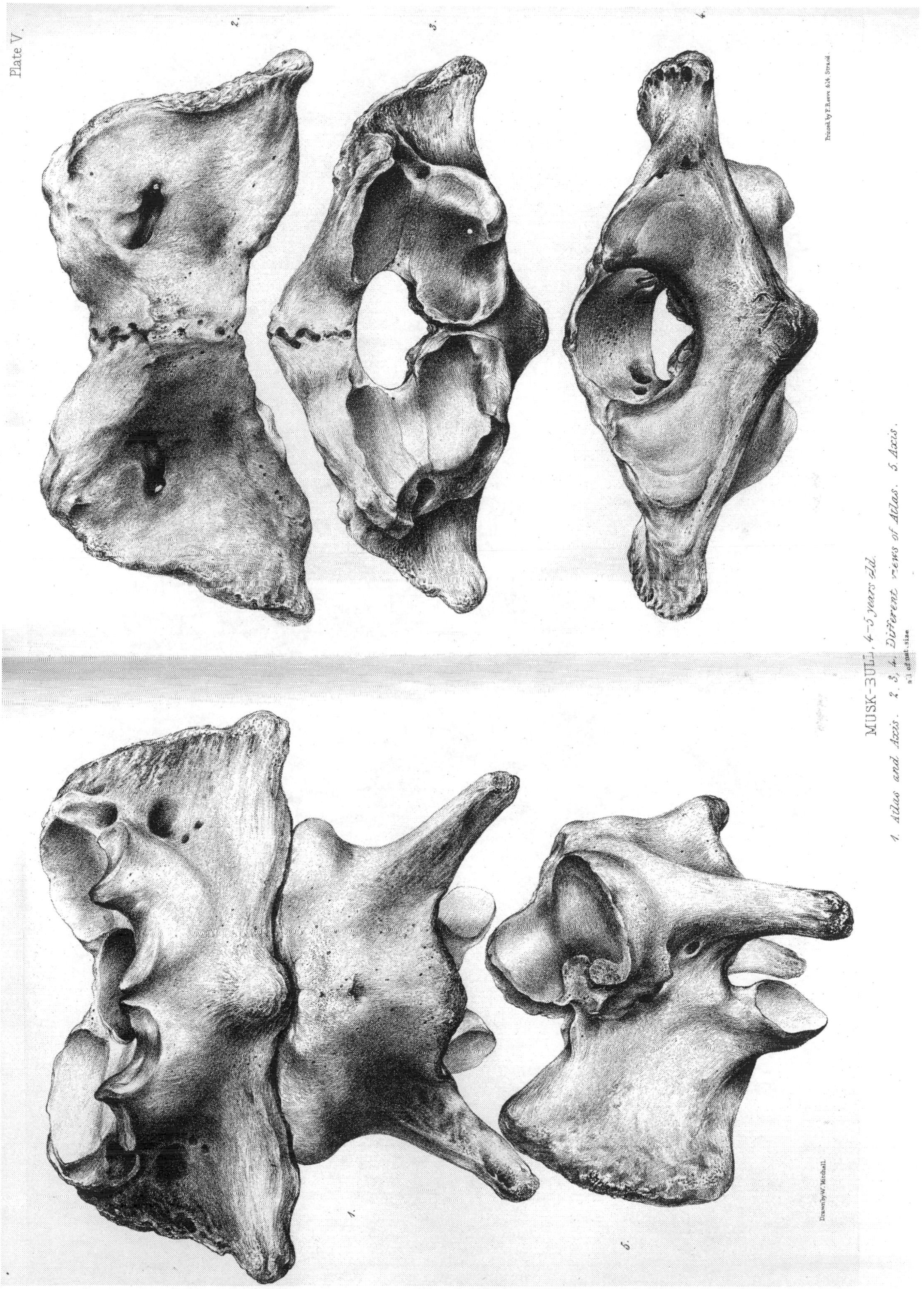

Plate V

Printed by F.Reeve &14. Strand

MUSK-BULL. 4–5 years 2d.

1. Atlas and Axis. 2.3,4. Different views of Atlas. 5. Axis.
a ⅓ nat. size

Drawn by W. Mitchall.

Plate VI.

Printed by J.Basire 6/8 Strand.

Drawn by P.Waterhouse Hawkins F.L.S.

1. AUROCHS. 2. D°   3 - 4 AMERICAN BISON.   5-6 FOSSIL.

All of the natural size.

Plate VI.

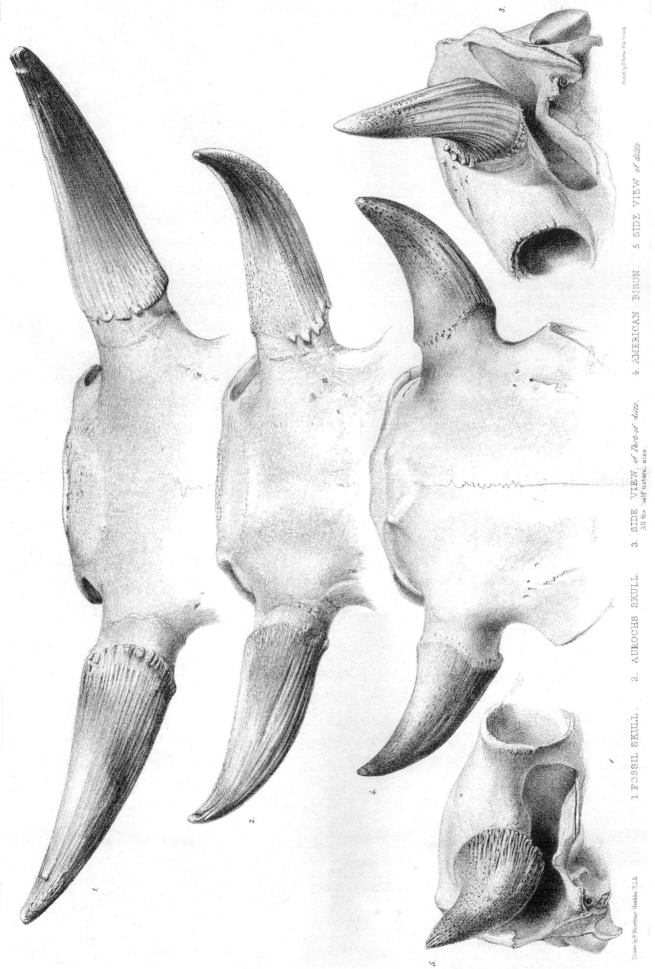

1 FOSSIL SKULL.   2. AUROCHS SKULL.   3. SIDE VIEW of *Pars of ditto*.   4. AMERICAN BISON   5 SIDE VIEW *of ditto*
All the *half natural size*.

Plate VIII.

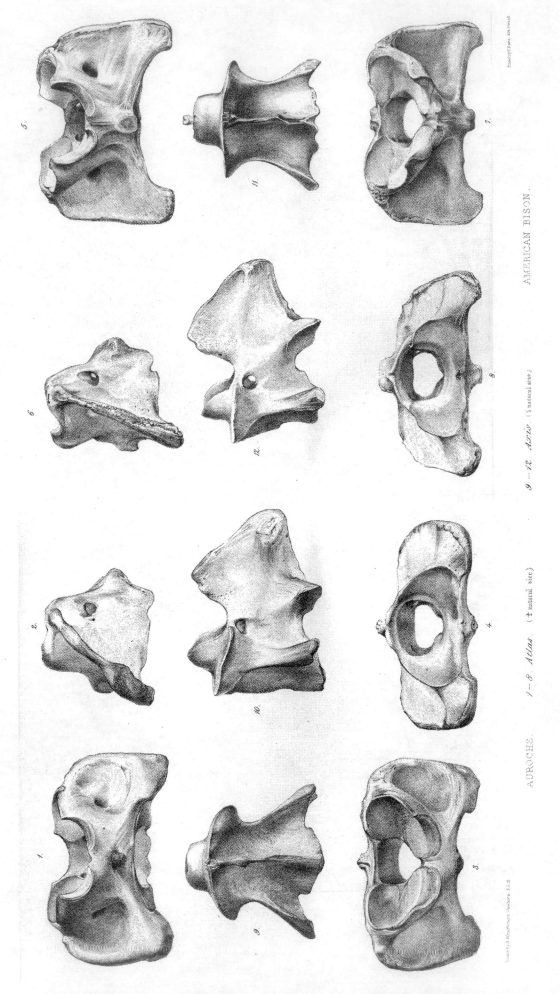

AMERICAN BISON.   9—12. *Aziz*. (½ natural size.)

1—8. *Atlas*. (¾ natural size.)

AUROCHS.

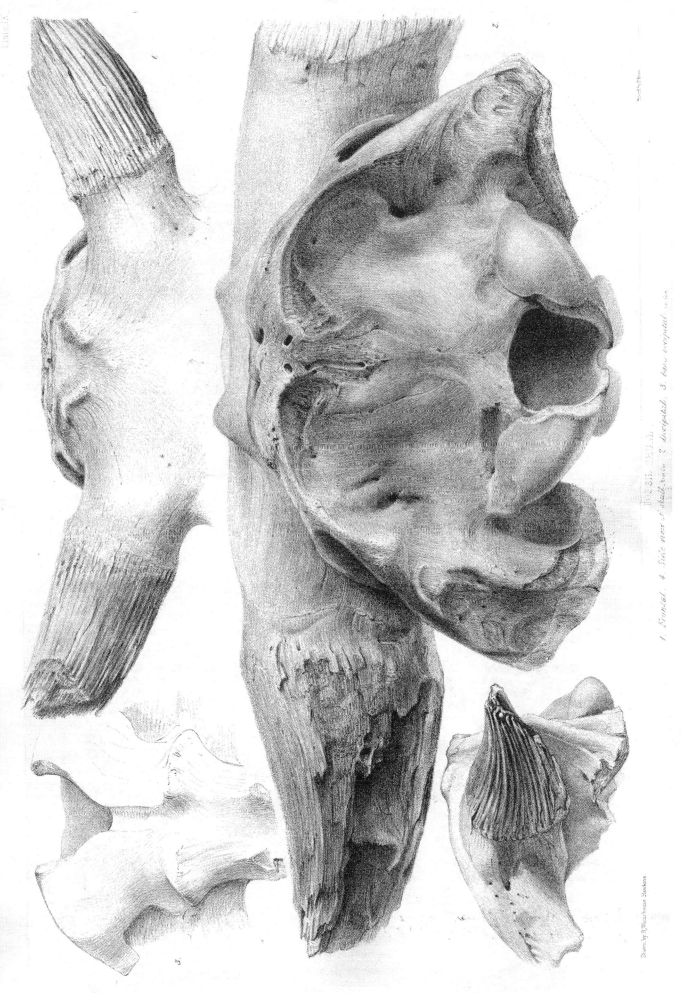

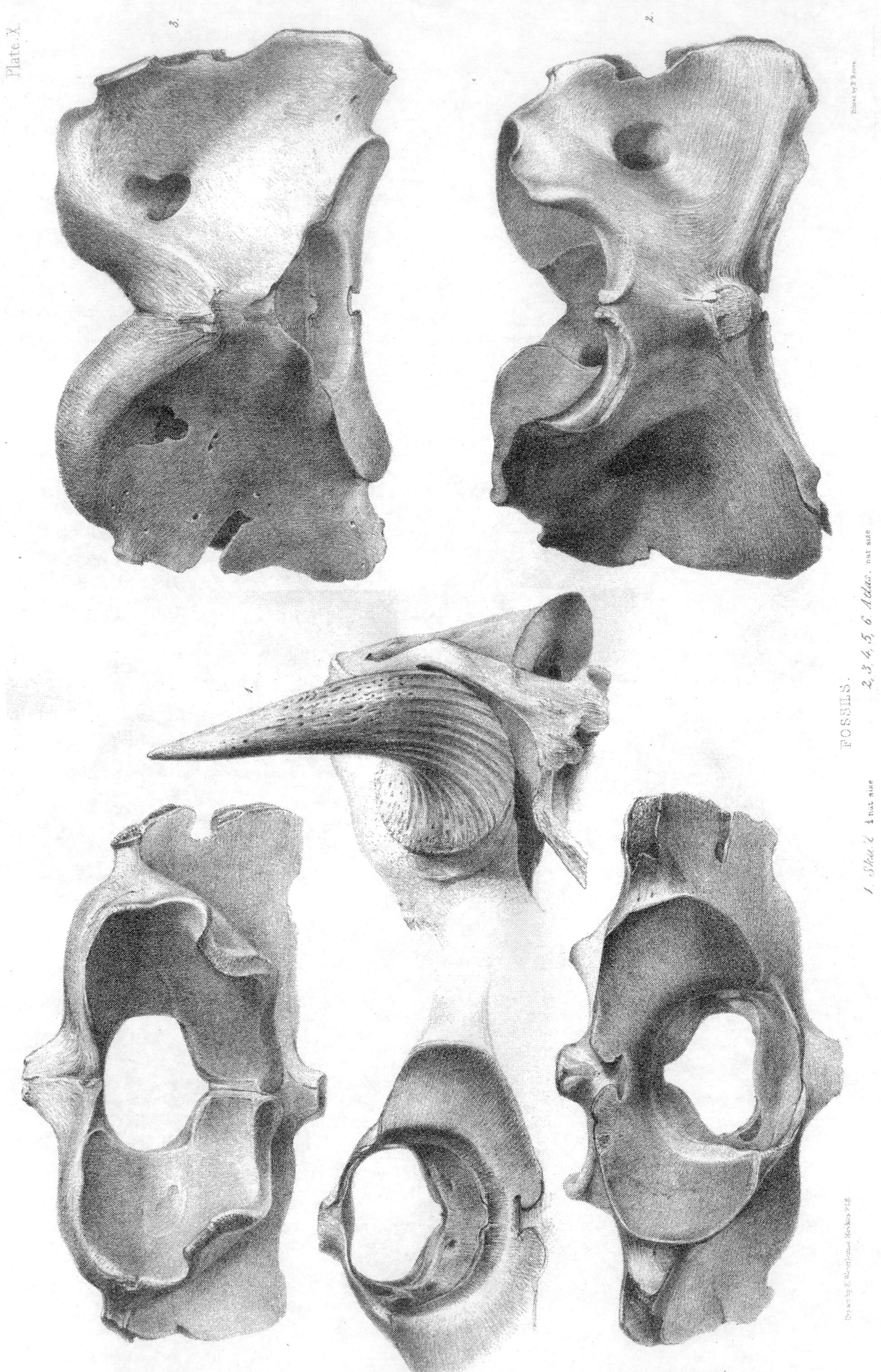

Plate X.

Drawn by R. Brown.

FOSSILS.

1. *Shell* ½ *nat. size*   2. 3. 4. 5. 6. *Atlas*. *nat. size*

Drawn by E. Rivollet and Hanhart, Printers, N.W.E.

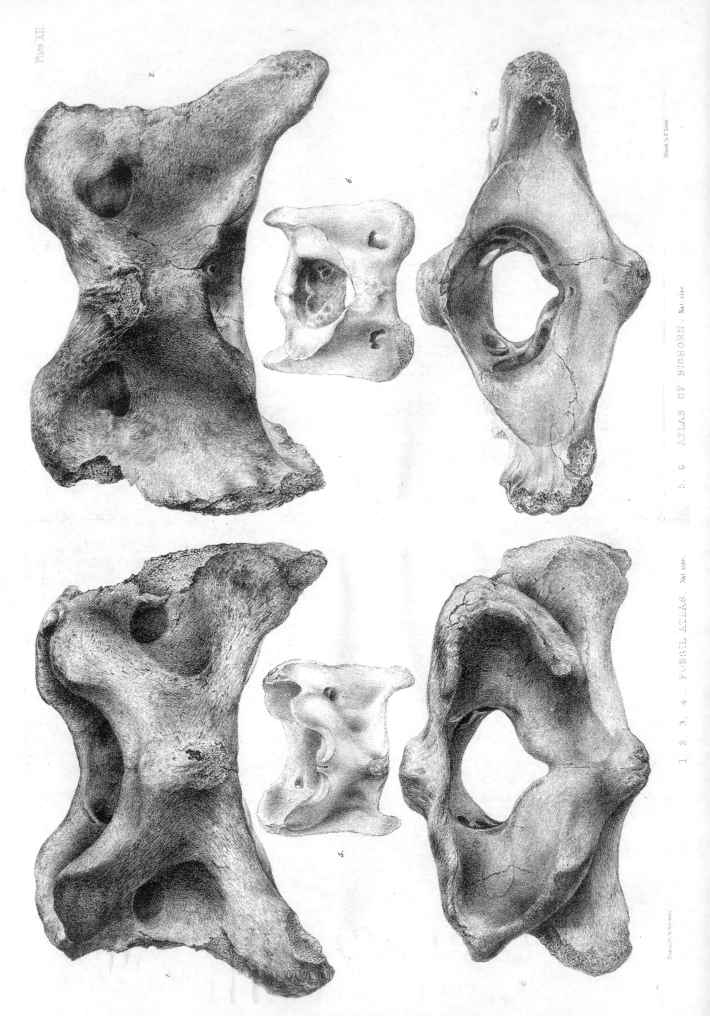

Plate XII

1. 2. 3. 4.. FOSSIL ATLAS. Nat.size. 5. 6. ATLAS OF BIGHORN. Nat.size.

Plate XIII

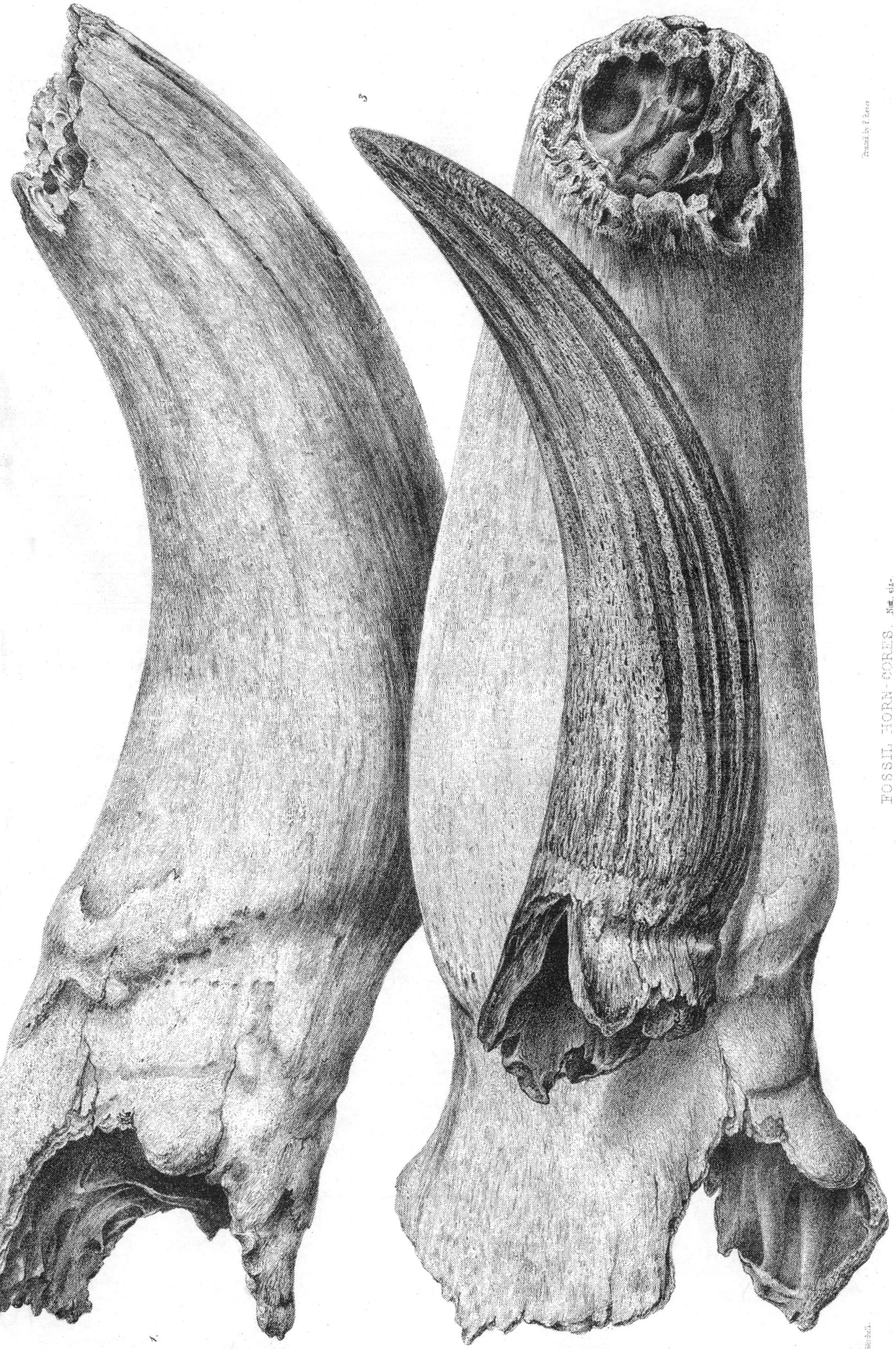

FOSSIL HORN-CORES.

The material originally positioned here is too large for reproduction in this reissue. A PDF can be downloaded from the web address given on page iv

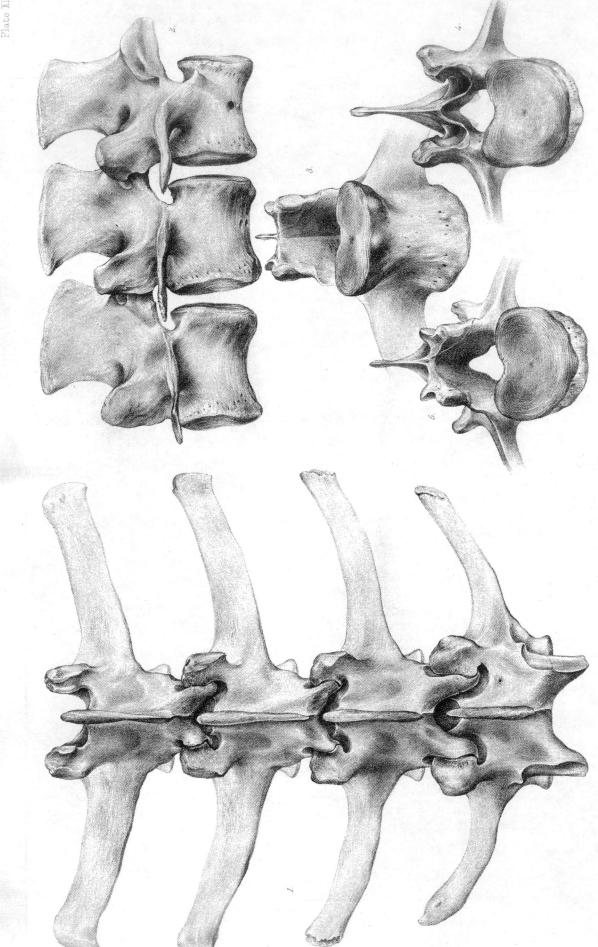

Plate XIV

MUSK COW.

1. Last four lumbars. 2. Last three lumbars. 3, 4, 5. Fourth Lumbar.

Plate XV

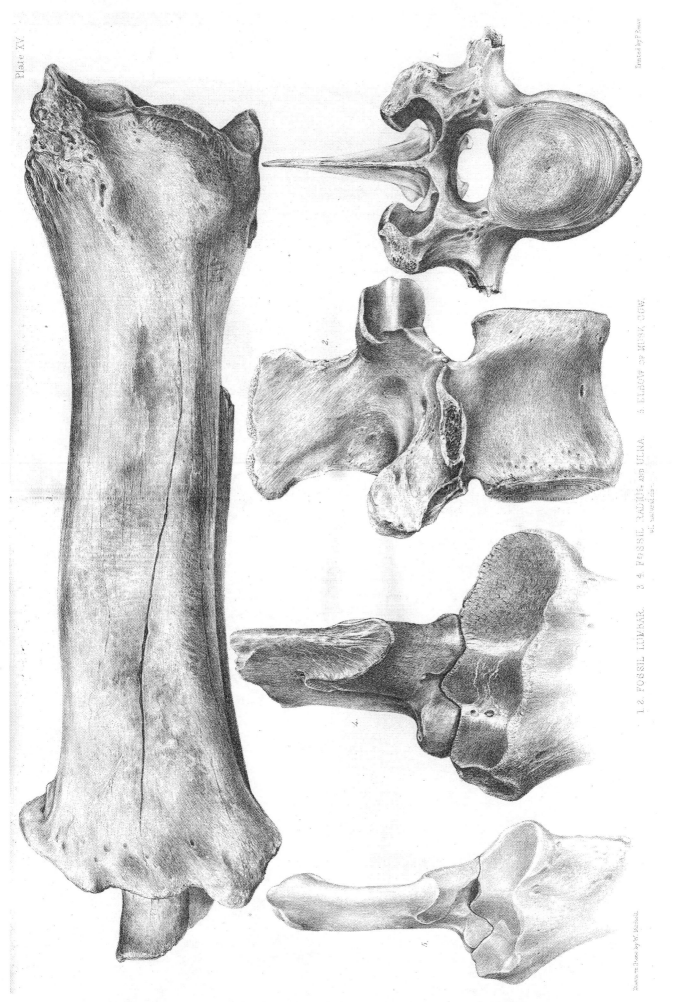

1. 2. FOSSIL LUMBAR.   3. 4. FOSSIL RADIUS and ULNA.   5. ELBOW of MUSK COW.

VERTEBRALS,

INCLUDING

FOSSIL MAMMALS.

THE ZOOLOGY

OF THE

VOYAGE OF H.M.S. HERALD,

UNDER THE COMMAND OF

CAPTAIN HENRY KELLETT, R.N., C.B.,

DURING THE YEARS 1845-51.

Published under the Authority of the Lords Commissioners of the Admiralty.

EDITED BY

PROFESSOR EDWARD FORBES, F.R.S.

VERTEBRALS,

INCLUDING

FOSSIL MAMMALS.

BY

SIR JOHN RICHARDSON, Knt., C.B., M.D., F.R.S.

LONDON:
LOVELL REEVE, 5, HENRIETTA STREET, COVENT GARDEN.

1854.

JOHN EDWARD TAYLOR, PRINTER,
LITTLE QUEEN STREET, LINCOLN'S INN FIELDS.

CONTENTS.

	PAGE.
On a pair of fossil shoulder-bones of the *Elephas Rupertianus*	101
On the osteology of the Muswa (*Alces Muswa*)	102
On the osteology of the Tuktu (*Cervus Tarandus*)	115
On the skeleton of an adult Musk-bull (*Ovibos moschatus*)	119
Further measurements of the skeleton of the European Aurochs-bull (*Bison Europæus*)	122
On the osteology of the Rocky Mountain Antelope (*Aplocerus montanus*)	131

LIST OF PLATES.

———◆———

PLATE XVI.

Lateral view of the skull and mandible of the Rocky Mountain Ram Antelope :—*nat. size.* (Page 131.)

PLATE XVII.

Fig. 1. Antinio-coronal view of the skull of the same male Rocky Mountain Antelope :—*nat. size.*

Fig. 2. Occiput of the same :—*nat. size.*

Fig. 3. Basilar view of the same :—*nat. size.* (Page 131.)

PLATE XVIII.

Fig. 1. Lateral view of the seven cervicals of the same skeleton of the Rocky Mountain Antelope :—*nat. size.*

Fig. 2. Sternal aspect of the same cervicals :—*nat. size.* (Page 131.)

PLATE XIX.

Fig. 1. Thenal aspect of the radius and ulna (*in situ*) of the same skeleton of the Rocky Mountain Antelope :—*nat. size.*

Fig. 2. Section of the same.

Fig. 3. Anconal aspect of the cannon-bone of the fore leg of the same skeleton, with a rudimentary fifth metacarpal attached :—*nat. size.*

Fig. 4. Section of the same.

Fig. 5. Popliteal aspect of the cannon-bone of the hind leg of the same skeleton of the Rocky Mountain Antelope :—*nat. size.* A circular flat sesamoid bone, which occupied the shaded depression at the proximal end of the bone, has been removed.

Fig. 6. Section of the same. (Page 131.)

PLATE XX.

Atlas, dentata, and third cervical of the " Muswa" (*Alces*), as seen on the sternal aspect :—*nat. size.* (Page 102.)

PLATE XXI.

Lateral view of the same three cervicals :—*nat. size.* (Page 102.)

PLATE XXII.

Fig. 1. Inial view of the skull of a male " Muswa," with its antlers :—*one-fourth* (*linear*) *of the nat. size.*

Fig. 2. Lateral view of the cannon-bone of the fore leg of a " Tuktu," or Barren-ground Reindeer (*Rangifer*), with the second and fourth metacarpals and their toes :—*nat. size.* (Page 102.)

PLATE XXIII.

Antinio-coronal view of the skull of a male "Tuktu," or Barren-ground Rein-deer :—*one-fourth* (*linear*) *of the nat. size.* (Page 117.)

PLATE XXIV.

Fig. 1. View of the glenoid cavity of the scapula, with its coracoid and spinous processes of the " Muswa" (*Alces*) :—*nat. size.*

Fig. 2–2. Thenal aspect of the radius and olecranon, with the interior of the elbow-joint of a " Buck Muswa:"—*nat. size.*

A section showing the position of the ulna at the middle of the shaft of the radius is given in the second figure.

Fig. 3. Thenal aspect of the cannon-bone of the fore leg (with the second and fifth metacarpals and their toes) of the same " Muswa:"—*nat. size.*

Fig. 4. Section near the middle of the bone. (Page 102.)

ZOOLOGY

OF THE

VOYAGE OF H.M.S. HERALD.

OSTEOLOGY—*Continued.*

REMARKS ON A PAIR OF FOSSIL SHOULDER-BONES OF AN ELEPHANT FOUND AT SWAN RIVER, IN RUPERT'S LAND.

THE first part of the Zoology of the Voyage of the Herald contains all the remarks that I originally intended to make on the fossil mammals of Eschscholtz Bay, or on the osteology of the living qua-drupeds of that region, introduced into this Work for the purpose of elucidating the extinct species. Subsequently however to the publication of that part I have, chiefly through the kindness and exer-tions of Dr. Rae, received skeletons of three mammals, which are important members of the existing Arctic Fauna, and also a pair of mutilated scapulæ of a fossil elephant. Taking these subjects in the order of the species mentioned in the preceding pages, the scapulæ come first under our notice; and although I have no intention of characterizing a species from the peculiarities of a single bone, it seems to be convenient for the sake of reference that these fragments should have a distinctive appellation, and I have adopted the geographical one of

ELEPHAS RUPERTIANUS.

No fossil bones of any kind had previously been discovered on the east side of the Rocky Mountains, to the north of the United States boundary (lat. 49°). These were found some years ago on the banks of the Swan River, near the western side of the basin of Lake Winipeg and its tributary Lake Winipegoosis. No information has reached me respecting the depth or extent of the alluvium in which they were imbedded, but they probably lay beyond the limits of the Silurian lime-stone deposit which constitutes the western shores of the above-named lakes. The scapulæ are right and left, and, being of the same size, were most likely members of the same individual. The broad ends of the blade in both are wanting, and most of the spinal ridge. On comparing what remains with scapulæ from Eschscholtz Bay, and with the homologous bone of the *Elephas primigenius* in the British Museum, I find that these fragments differ in possessing a well-marked depression between the humeral end of the ridge and the anconal edge of the bone. This depression is 4½ inches long,

P

and has an abrupt edge, upwards of an inch deep, next the spinal ridge, but gradually passes into the flat bone in other directions. No such depression exists in the shoulder-blade of the Mastodon; neither, as has been said above, does the Mammoth's scapula show a depression with a steep edge, though at the same part there is a scarcely perceptible hollow, not more evident than in the shoulder-bone of the existing Asiatic species. There is however a scapula in the British Museum from Himalaya, figured by Dr. Falconer, which has an evident depression at this part, but without the abrupt edge.

E. Rupertianus.

The length of the glenoid cavity is	9·1 inches.
Its breadth	5·5
The space between the depression and the acromial edge of the bone where narrowest is	4·0
Length of the largest fragment	17·5

ALCES MUSWA. *The Moose-deer.*

Muswa of the Cree Indians (pronounced *mooswăw*). (Plates XX., XXI., XXII. fig. 1, and XXIV.)

When I drew up the account contained in the preceding pages of the fossil bones disinterred from the cliffs of Eschscholtz Bay, I had not the advantage of comparing them with the bones of the Moose-deer, an animal exceeding the domestic ox in stature, and frequently equalling it in the bulk and weight of its carcase. Through the kindness of Dr. Rae and some of his brother officers in the Hudson's Bay Company's service, I have been furnished with a complete skeleton of the Moose-deer, and shall proceed to give some account of its peculiarities in the following pages.

Cranium
of a Buck of the age of four years, or thereabouts. (Plate XXII. fig. 1.)

In the abbreviation of the nasals, the prolongation of the maxillaries and premaxillaries, and in the elongated parietals extended in the same plane with the face, the *Muswa*, or Moose-deer, differs from the other *Cervidæ, Bovidæ*, or *Capridæ*. The summit of the skull on its coronal aspect is formed by the median elevation of a rounded ridge, which crosses from the basis of one antler to that of the other, and is cut at right angles by the sagittal suture, whose raised edges constitute the summit in question. Into the composition of the transverse ridge the parietals enter, but it is constructed chiefly by the frontal; and at its lateral termination each way, and on its inial aspect, there is a short conical protuberance belonging to both bones. The distance from the apex of one of these protuberances to that of the other is 4 inches, and the spaces between them and the orbital plates on their respective sides are occupied by the swelling bases of those processes, which, having a lateral direction, with a slight inclination coronad, give origin and support to the antlers. Close to the prominent basal ring of the antler the process has a circumference of 6½ inches.

In the middle of its length the *frontal* is bent inwards towards the encephalon, as if it had received a violent blow when in a plastic state, and the hollow is divided longitudinally by the raised edges of the sagittal suture. A similar incurvature, but not to the same extent, exists in the frontal of the Rein-deer. Opposite to the antinial border of the depression in the Muswa the supra-orbitar foramina perforate the orbitar plates, which are thinner and less prominent than those of the Musk-ox.

As it is in the development of the *parietals* that the peculiarly elongated form of the posterior calvarium depends, so we find that these bones differ in form from their homologues in other ruminants. Conjointly they are bounded next the frontals by a straight transverse edge, having a very concave occipital one opposite to it, and laterally on each side by less concave edges, which the squamosals overlap. In forming the posterior transverse boundary of the frontal, the coronal suture does not bend antiniad in its median portion, as it does in the Rein-deer cranium. The sides of the encephalon are embraced in the usual manner by the lateral antinial parietal processes, which interpose between the squamosals and the orbital plates of the frontal; but the points of these processes do not reach the orbito-sphenoids as in the Rein-deer, owing to the interposition of the expanded tips of the ali-sphenoids. There may however be some variety in this respect in different individuals, as the two sides in our specimen do not correspond. On the left side a pretty large sesamoid bone is intercepted by the suture, and the relations of the parts are thereby somewhat obscured.

On the sides, the parietals are bounded by the widely-arched squamose suture, and their inial ends are deeply indented by the super-occipital. In its turn the latter bone receives the acute lateral inial angles of the parietals into a smaller notch. Plate XXII. does not exhibit the occipital suture, which, on that point of view, is concealed by the crest of the bone. Complete coalescence having taken place between the parietals and inter-parietals in our specimen, the form of the latter cannot be made out, and the only remaining vestige of the sagittal suture in its course between the parietals is a slightly raised line. There is also a curved line on each parietal, concentric with the squamose suture, marking the attachment of the temporal muscle.

Except in the camel, none of the genera allied to *Cervidæ* and *Bovidæ* present an example of so prominent an occipital crest as this deer. Rounded in outline laterally, the crest is notched mesially in the usual site of the occipital spine; and beneath its margin, the super-occipital plate is concave and rough for the attachment of muscles, but the ex-occipitals have an inial slope as they approach the *foramen magnum*, whose nearest border however still falls an inch antiniad of the overhanging margin of the crest.

The lateral edges of the condyles are acute and free, without the slighest tendency to form an accessory trochlea as in the musk-ox; and a defined transverse line on each condyle indicates the meeting of its two articular surfaces.

No trace of a suture remains to mark the union of the basi-occipital with the basi-sphenoid, which have conjointly a truncated conical outline and a very convex surface, culminating in the basi-occipital in a prominent median line. This bone has also projecting shoulders, against which the atlas abuts in nutation, and a rough spot, conical in outline, exists on each side at the junction of the bones, where, in most of the *Bovidæ*, there is a prominence. The ali-sphenoid has also coalesced extensively with the basi-sphenoid.

About one-third of the distance between the frontals and the extreme tip of the premaxillaries is occupied by the nasals. These bones, representing the divided neural spine of the rhinencephalic vertebra, are subject to greater variety than the centrum, and, judging from the discrepancies that occur in the few crania of the elk that I have had an opportunity of examining, the variations in that species are more frequent than in other deer. In our specimen the right nasal is shorter than the left one, the transverse facial suture being rendered irregular by the antinial projection of a short angular process of the frontal on that side. The cranium of one American elk in the Museum of the College of Surgeons in London, agrees with ours in this respect, while in another in the same collection there is a triangular mesial bone interposed between the nasals for nearly their whole length, this skull having in fact three nasals. In the skull of a European elk, also in that collection, this intercalated bone is smaller and more irregular in shape, and looks like a large sesamoid bone. The nasals of our specimen differ from all the three just mentioned, in having a suborbicular bone, half an inch in diameter, united to the antinial extremity of each of them by suture. With respect to the general form of the nasals of the Muswa, they are rather broad in a lateral direction, and much arched, not regularly, but abruptly bent, so that each bone has a defined lateral as well as a coronal surface.

By the abbreviation of the nasals, much of the interior structure of the nostrils is exposed in the skeleton. A very complicated ethmoturbinal occupies the angle between the frontal, maxillary and lachrymal, and nearly equals the latter in the area of the portion which enters into the composition of the face. It exhibits three deep longitudinal cells, separated from each other by thin bony partitions, and a long papyraceous spongy plate of the bone in the cavity of the nose underlies the nasal throughout, arching over the proper turbinal, which is also enormously developed, and projects antiniad far beyond the nasals. About one-fourth part of the border of the orbits is formed by the lachrymal, which on its facial aspect has an acutely triangular shape, its apex entering a notch of the maxillary.

The lateral limb of the premaxillary is received as usual into a groove of the maxillary, but its point falls short of reaching the nasal bone, the intermediate corner being filled by a triangular piece of bone, in great part coalescent with the maxillary, but which in the young animal was probably a facial plate of the turbinal. Considering the size of the skull, the antinial extremities of the premaxillaries are slender and narrow. The foramen formed between their limbs and the maxillary is $4\frac{1}{2}$ inches long. The breadth of the face at the most protuberant parts of the maxillaries is less in the Muswa than in the European elk, as I ascertained by comparing the crania in the College of Surgeons, and as Mr. Quekett had previously observed. Though I am inclined to consider the American elk as specifically distinct from the European animal, I have not thought it necessary to use any other distinctive name than the trivial one of *Muswa*.

Between the rows of the molars the palate is moderately concave transversely, and slightly convex longitudinally. It measures $2\frac{1}{2}$ inches across at the first premolars, and $3\frac{1}{2}$ inches at the last molars. In length the entire molar series is 6 inches, while the distance from the first premolar to the tip of the premaxillary is $9\frac{1}{4}$ inches. Anterior to the premolars the palate is narrowed to an almost semicircular channel, bounded by thin ridges of the maxillary, but widens and flattens again where it is formed by the premaxillaries.

Teeth of the Muswa.

Incisors, $\frac{0-0}{3-3}$; Canines, $\frac{0-0}{1-1}$; Premolars, $\frac{3-3}{3-3}$; Molars, $\frac{3-3}{3-3}$; Total, $\frac{6-6}{10-10}$.

A complete saw is formed by the very acute cusps of the molars. Each of the maxillary molars has a minute projection or denticle on the mesial aspect of its posterior lobe, and the lateral surfaces of both molars and premolars have a strong inclination mesiad. The inclination in the opposite direction of the mesial surfaces of the mandibular molars is not so great, and it is only the first mandibular molar that has the accessory denticle, situated of course on its lateral aspect. No more effective instruments could be devised for cutting the flexible willow-twigs on which this animal browzes than its molar series of teeth, and they retain all their sharpness in use. The incisors are also acute-edged, and differ little from one another in size, the inciform canine however, and the tooth which adjoins it, being a little narrower than the middle pair. There are no vestiges of the upper canine which exists at the point of the maxillary in both male and female rein-deer.

Antlers of the Muswa.

These extraordinary deciduous growths take their origin, as stated above, in the lateral processes of the frontal, which project about an inch and a half from the sides of the skull. A prominent ring, rough with blunt wart-like projections, marks the commencement of the antler, and is finally, by obliterating the nourishing arteries, the cause of its fall. Beyond the ring there is a stem, round at the very beginning, but becoming speedily more and more compressed, until it expands into a large palmated plate, which is so curved that its inial and antinial halves make an angle with each other of 130 degrees, the curve being however gradual rather than sharp, and having its apex situated about three inches iniad of the axis of the stem, thus dividing the palm into two unequal planes. The antinial expansion rises coronad, with a slight inclination antiniad, but not so much as to cause it to pass the line of the frontal borders of the orbits, while the direction of the inial and larger portion is, after the curvature becomes complete, nearly directly iniad. In Plate XXII. fig. 1, the foreshortening of this part has not been so skilfully executed by the artist as to convey a very exact idea of its extent and direction. It is proper to mention that, the drawing not having been reversed on the stone, the right antler is represented as the left one, and the left as the right, in the figure. One antler presents eleven short snags and the other fourteen, with sinuses of various depth between them, but these incurvatures are on the whole considerably less deep than in any of the antlers of the European elk preserved in the College of Surgeons or British Museum. The longest snags are the pair that stand nearly in the axis of the stems, and their tips are 22 inches from the sagittal suture, or 40 from each other. From the antinial snag to the last inial one, the distance is 27 inches in one antler, and nearly two inches less in the other. The front snags of the two antlers are 10½ inches apart, and the two most inial ones nearly 28 inches. The weight of the antlers with the skull and mandible is nearly 24 lbs. avoirdupois, and, from the large development of the fleshy and flexible nose and lips, the head of the living animal must weigh very much more, and require powerful muscles for its support.

At page 20, No. 114, mention is made of a fragment of a fossil moose-deer skull, which, on a close comparison with the same part of the recent animal, presents some differences. It belonged in the first place to a larger individual, the distance from the sagittal suture to the basal ring of the antler being half an inch greater. The edges of the suture just mentioned do not project in the same degree as in the recent skull, and the stem of the antler is larger, longer, and less compressed. A fragment of a fossil antler, No. 198, mentioned also at page 20, cannot be referred to any part of an antler constructed like that of our recent skeleton, but its size and palmature point to its being a portion of the antler of an elk, and, from its conformity in condition to No. 114, probably a relic of the same individual, having been found in the same locality.

Dimensions of the Skull of the Muswa (by callipers).

Length of the skull from the antinial end of the premaxillary to the occipital ridge .	21·7 inches.
Length from the same point to the transverse suture at the root of the nasals . .	13·7
Length from the transverse suture to the occipital ridge, mesial line . .	9·7
Breadth at the orbits, inial edges	8·7

Breadth at the orbits, basal edges . . . . 8·5 inches.
Breadth at ditto, antinial edges, between the projecting corners of the lachrymals . . 7·2
Breadth of frontals between the orbits and antlers . . . . . . 7·8
Breadth at the widest part of the squamosals, adjoining the paroccipitals . . 6·5
Breadth between the outsides of the zygomatic arches . . . . . . 8·1
Distance between the lateral edge of one condyle to that of the other on the inial aspect 3·7
Distance between the occipital crest and the basilar edge of the foramen magnum on the
    mesial line . . . . 5·2
Distance between the basilar edge of the foramen magnum and the antinial end of the
    basi-sphenoid . . . . 3·8
Breadth of the shoulders of the basi-occipital . . . . . . 2·4
Distance between the points of the exoccipital spinous processes . . . 3·7
Greatest width of the opening to the posterior nostrils . . . . 1·5
Width of the maxillæ at the fangs of the first and second true molars . . . 6·0
Distance between the roots of the incisors and the posterior border of the mandibular
    condyle . . . . 19·2
Distance from the same point to the inial curve at the angle of the mandible . . 18·7
Rise of the coronoid process above the surface of the condyle . . . 2·7
Chord of the premolar and molar series . . . . 6·4
Distance from the roots of the medial pair of incisors to the first premolar . . 8·0
Distance from the last molar to the inial curve of the jaw . . . . 4·4

Vertebræ of the Muswa.

Measured over their neural spines the seven *cervicals* have a length of 12 inches, and following the curve of their centra on the sternal aspect 23 inches. Less massy and weighty, and inferior in all its dimensions to its homologue in the musk-bull, the *atlas* agrees in size and very closely in form with that of the Alderney cow. On comparing this cervical with the corresponding one of the domestic ox, the principal differences elicited are the greater depth and abruptness of the depressions in the lateral processes beneath the arterial foramina, the greater thickness of the edges of these processes opposite to the sacral limits of the depressions, and the more direct projection sacrad of the distal ends of the same processes. In the domestic ox these distal ends diverge more, and are thickest at the point. The antarticulating surface of the centrum of the muswa's atlas is also more rounded off sternad, indicating a greater play against the shoulders of the basi-occipital in the nutatory motions of the head; and there is a greater prominence of the comparatively slender hypapophysial knob, as is well shown in the side view given in Plate XXI., which exhibits the direction of the lateral process, and the manner in which it differs in the thickness of its edge from the same part in the domestic ox. Plate XX. gives a sternal view of the atlas, and may be consulted with regard to other peculiarities, but the general resemblance of the bone to its bovine homologue is such that comparison rather than descriptions must be had recourse to for discrimination. A shallow notch separates the neural and sternal antarticulating surfaces. This atlas is rather too small for the condyles of the skull described above.

Within the neural canal the centrum presents a raised median line, excepting on the smooth surface on which the odontoid process plays; and there is a considerable depression on the proximal edge of the canal in the corner between the centrum and arterial foramina. This place is rough but not excavated in the ox, though there is a hollow in the wall of the canal a little more sacrad.

ATLAS.

Extreme breadth between the lateral edges of the atlas (at the distal third) . . .	5·6 inches.
Breadth at the proximal end of the lateral processes	4·5
Distance between the outer borders of the distal articulating sufaces . . .	3·8
Transverse diameter of the proximal cup for the reception of the condyles . .	3·6
Lateral axis of the neural canal	1·6
Sterno-neural ditto	1·3
Distance between the peripheral points of the neural and hypapophysial knobs . .	3·5

In general form the *dentata* differs very little indeed from the same segment of the vertebral column in the domestic ox. Its hypapophysial ridge however is not lengthened so much sacrad, and consequently the postarticulating cup is less oblique. The diapleurapophyses also curve more sacrad, and project less laterally. Differences moreover may be discovered, by comparison, in the curves of the articulating surfaces of the zygapophyses, and also in the neural surface of the centrum, which, in the muswa, supports an acute median ridge within the canal, while in the domestic ox the same part is quite flat. The neural spines of the two species are almost exactly alike, and the resemblance between their proximal articulating surfaces is also very close. On the back of the zygapophysis there is a short conical and rather acute eminence, which does not exist in the *dentata* of the ox.

DENTATA.

Transverse diameter of the antarticulating surface	3·7 inches.
Transverse diameter of the postarticulating surface	1·7
Sterno-dorsal diameter of ditto	1·7
Height of the odontoid process on its sternal aspect	0·9
Length of the centrum on its sternal side, excluding the articulations . . .	3·7
Its width where narrowest atlantad of the root of the diapophyses . . .	2·0
Distance from the sternal side of the centrum, atlantal edge, to the atlantal corner of the neural spine	3·8
Distance from the sternal side of the centrum, sacral edge, to the sacral end of the neural spine	5·2
Length of the crest of the neural spine between its atlantal and sacral corners . .	3·6
Length of the neural arch or base of the neural spine	3·1
Distance from the outer side of one zygapophysis to that of the other . . .	3·2

The *third cervical* (Plates XX., XXI.) differs from its homologue in the ox in the comparative weakness of its transverse process. Its distal end, which in the succeeding vertebra separates laterally as the diapophysis, is much less ridged on the back, and the hypapophysis is more acute and less prominent, and projects less sacrad; the neural spine also tapers more; but all these differences are chiefly in degree, and the general form of this vertebra in the two species is much the same.

In nearly the same points, similar differences exist between the *fourth cervicals* of the muswa and domestic ox, with this addition, that the deep cavity which lies between the bases of the diapophysis and parapophysis in this vertebra of the domestic ox, does not exist in the same bone of the muswa, the two processes in the latter animal being coalescent much as they are in third cervical. The neural spine also has a smaller inclination atlantad than in the *Bovidæ*.

The *fifth cervical* however differs more from its homologue in the common ox, by the marked abbreviation of the parapophysis and diapophysis, more especially of the latter. Its neural spine also is less inclined atlantad.

By the shortness of the diapophyses and the direction of the neural spine, the *sixth cervical* of the muswa may be readily distinguished from the same bone of the common ox. In the latter the neural spine makes an angle of about 60° with the axis of the neural canal, while in the muswa the angle is more nearly a right one.

	CERVICALS.			
	THIRD.	FOURTH.	FIFTH.	SIXTH.
Breadth of the antarticular surface of the centrum . .	1·4 in.	1·5 in.	1·5 in.	1·3 in.
Sterno-dorsal diameter of ditto	1·6	1·7	1·7	1·7
Breadth of the postarticular cup of the centrum . .	1·8	1·8	1·8	1·8
Sterno-dorsal diameter of ditto	1·9	1·9	2·0	2·0
Length of the centrum on its sternal aspect, excluding the articulations	2·7	2·6	2·7	2·2
Height from the tip of the hypapophysial knob to the summit of the neural spine	5·1	5·8	6·4	6·4
Breadth from the tip of one diapophysis to the tip of the other, minus the epiphysis	4·4	4·8	4·3	4·2
Distance from the lateral corner of one parapophysis to that of the other, minus the narrow epiphysis . .	2·2	3·2	3·8	3·0
Distance from the most lateral part of one prozygapophysis to that of the other	3·4	3·4	3·4	3·5
Distance from the most lateral part of one zygapophysis to that of the other	3·2	3·2	3·5	3·5
Length of the neural spine above the crown of the arch, at its proximal edge	2·0	2·0	2·6	3·0
Transverse distance between the outer wall of one arterial foramen and that of the other	2·4	2·6	3·0	3·1

A greater discrepancy of general form exists between the *seventh cervical* of the Muswa and its bovine homologue, than is exhibited by any of the preceding cervicals. Its centrum is more abbreviated than the same part of the domestic ox, its diapophysis much shorter, simpler, more removed from the prozygapophysis, and connected to the centrum by a broader plate of bone. The pleurapophysial epiphysis is very thin, the neural arch has a narrower crown, and the neural spine is longer and inclines rather sacrad, while in the common ox its inclination is slightly atlantad.

	SEVENTH CERVICAL.
Breadth of the antarticular surface of the centrum	1·5 inches.
Sterno-neural diameter of ditto	1·9
Breadth of the postarticular cup of the centrum, including cups for the heads of the ribs	2·7
Sterno-dorsal diameter of ditto	1·9
Length of the centrum on its sternal aspect, excluding the articulations . . .	1·4
Distance from the tip of one diapophysis to that of the other, including the pleurapophysis	4·3
Distance from the most lateral part of one prozygapophysis to that of the other . .	3·7
Distance from the most lateral part of one zygapophysis to that of the other . .	3·1
Length of the neural spine above the crown of the arch at its proximal edge . .	5·7
Length of ditto above ditto at its distal edge	5·4
Transverse distance between the edges of bone that connect the prozygapophyses and diapophyses	2·3

The *dorsals* are thirteen in number. Measured over their neural spines their length is 25½ inches, and along the sternal surfaces of their centra 26½ inches. Though the second and third spines are absolutely the longest of the series, the summit of the back is formed by the fourth and fifth. The first and fifth equal each other in height. In general form and relative development of their metapophyses the dorsals of the muswa agree closely with those of the domestic ox, and their sizes in our specimen correspond with those of an Alderney cow of the ordinary stature. The general resemblance of the segments of the spinal column of the two species is such that a careful comparison is requisite to discover their discrepancies, and mere description will scarcely suffice for recognition. In the muswa the neural processes are taller, and as thick, but the neurapophyses are less full and strong, and the sockets which receive the heads of the ribs are generally larger. The acquisition of this skeleton induced me to re-examine the fossil vertebræ referred in the preceding pages to different species of *Bovidæ*, but a close comparison has not excited doubts of the correctness of those references. Number 132, (described at p. 38,) the smallest of the bovine vertebræ received from Eschscholtz Bay, is larger than a dorsal of our muswa taken from the same part of the series, and presents differences in its processes.

With respect to the development of the metapophyses in the muswa, a round tubercular thickening of the back of the diapophysis of the second dorsal, becomes a small conical projection on the fourth, and attains its greatest prominence on the eleventh, and has an antlantad direction. In the twelfth the prozygapophysial articulating surface changes from the front of the neural arch to the medial aspect of the metapophysial tubercle, a point of the tubercle however having still a small projection atlantad. In the thirteenth the projection disappears, and the metapophysis is a mere thickening of the back of the prozygapophysis.

DORSALS OF THE MUSWA.

	FIRST.	FOURTH.	EIGHTH.	TWELFTH.
Length of the centra, excluding the articulations	1·5 in.	1·8 in.	1·9 in.	2·0 in.
Width of their antarticulations, including the cups for the heads of the ribs	2·5	1·7	1·8	1·9 .
Sterno-neural diameter of their antarticulation	1·8	1·5	1·4	1·4
Transverse diameter of the centra in the interval between the proximal and distal cups for the ribs	2·1	1·5	1·3	1·5
Transverse distance from the most lateral part of one diapophysis to that of the other	4·3	3·5	3·3	3·6
Length of the neural spine above the crown of the neural arch on its proximal edge	8·0	10·0	5·5	2·1

The six *lumbars* measure over their neural spines 14·5 inches, and over their centra, allowing space for intervertebral substance, 15·5 inches. Their resemblance to the homologous bones of the domestic ox is still stronger than that of the dorsals. The centra have the same lengths with those of an Alderney cow, but they are rather fuller on the sides, though their hypapophysial ridges are equally distinct.

MUSWA LUMBARS.

	FIRST.	FOURTH.	SIXTH.
Length of the centra, excluding the articulations	2·0 in.	2·4 in.	2·1 in.
Width of their antarticulations	1·6	2·0	2·1
Transverse diameter of the bone at the interval between the prozygapophysis and diapophysis	2·0	2·3	3·2

Q

	MUSWA LUMBARS.		
	FIRST.	FOURTH.	SIXTH.
Transverse distance from the tip of one diapleurapophysis to that of the other, excluding the thin epiphyses, which have dropped off in maceration	6·5 in.	9·0 in.	8·4 in.
Distance from the sternal surfaces of the centrum to the tip of the neural spine, excluding the epiphysis	4·3	4·5	4·1
Transverse distance from the most lateral part of one metapophysial tubercle to that of the other	2·2	2·7	3·2
Transverse distance between the outsides of the zygapophyses . .	1·3	1·6	2·3
Length from the proximal part of the prozygapophysis to the distal part of the zygapophysis	3·3	3·8	3·4

Sacrals.—The sacrum is composed of five pieces, none of whose centra are coalescent, though the summits of the neural spines of the first three are. There are six *coccygeals*.

	SACRALS.
Length of the chord of the five sacrals on their pubal aspect	9·2 inches.
Greatest transverse diameter of the lateral processes of the *first sacral* at its proximal end	6·5
Transverse diameter of the proximal articulation of the centrum of the *first sacral* .	3·0
Distance from one prozygapophysis to the other of the *first sacral* . . .	2·0
Length of the centrum of the *first sacral*	2·1
Length of the centrum of the *second sacral*	2·0
Length of the centrum of the *third sacral*	1·7
Length of the centrum of the *fourth sacral*	1·6
Length of the centrum of the *fifth sacral*	1·6
Breadth of the lateral processes of the *fifth sacral*	2·2
Length of the *six coccygeals*	8·0

Pelvis of the Muswa.

The pelvis strongly resembles that of the domestic ox in its general form, but a comparison of one with the other exhibits well-marked differences. The ramus of the ilium is much thicker and rounder on its sternal aspect than that of the ox, and the sternal angle of the expansion of the bone is less prominent. When contrasted with the pelvis of an Alderney cow of much lower stature and of less weight than a buck muswa, the pelvis of our specimen is found to have smaller dimensions generally. Its sternal outlet is shorter in its sacro-pubal axis, and narrower transversely in a still greater degree. The notch between the ramus of the ilium and sacrum is also less wide than in the ox, and there is only one notch in the brim of the acetabulum, viz. the one directed towards the foramen ovale. The other notch which exists in the ox between the pubal and ischiatic portions of the acetabulum is filled up in the muswa.

	PELVIS.
Distance from the sternal angle of the crest of one ilium to that of the other . .	13·2 inches.
Transverse distance from the same angle to the neural angle of the same ilium, or length of the sterno-neural chord of the crest	7·8
Sterno-neural diameter of the iliac shaft at its slenderest place	1·5
Sterno-neural diameter of the acetabulum	2·1
Atlanto-sacral ditto	2·2

Transverse distance from the apex of one conical process, situated near the neural angle of the ischium to that of the corresponding process of the other ischium . . 8·0 inches.

Transverse distance between the neural or spinous angles of the ischia, being the width of that part of the pelvis . . . . 4·4

Transverse diameter of the pelvis in front between the rami of the ilia . . . . 4·7

Length of the foramen ovale . . . . 3·5

Sterno-neural diameter of ditto . . . . 2·2

Length of the pelvis from the most proximal part of the crest of the ilium to the tuberosity of the ischium . . . . 18·6

Distance from the dorsal angle of the ischium to the summit of its lateral conical process 3·1

Sterno-neural diameter of the dorsal ramus of the ischium . . . . . . 1·7

Distance between the tuberosity of the ischium and the nearest part of the brim of the acetabulum . . . . 7·3

Distance between the atlantal edge of the acetabulum and the sternal corner of the iliac crest . . . . 8·5

Length of the proximal ramus of the pubal, between the acetabulum and symphysis . 3·3

Atlantal Extremity of the Muswa.

The *scapula* is large, and exceeds in its dimensions its homologue in the Musk-ox or Alderney cow. Its blade is as large as that of a full-sized domestic ox, but its glenoid cavity is smaller. The form of this cavity is represented in Plate XXIV. fig. 1, of the full size.

Length between the radial edge of the glenoid cavity and the atlantal corner of the dorsal edge of the bone (excluding the appendix, which is chiefly cartilaginous) . 16·0 inches.

Length of the dorsal edge of the bone . . . . 10·3

Length between the apex of the coracoid tubercle and the distal or anconal edge of the glenoid cavity . . . . 3·2

Width of the narrowest part of the neck of the bone (its radio-anconal diameter) . . 2·2

Height of the front of the spine of the scapula . . . . 2·7

Long axis of the glenoid cavity . . . . 2·4

Transverse axis of ditto . . . . 2·2

Humerus.—This bone shows less variety of form in the ruminants than most other parts of the skeleton, for though a minute comparison elicits differences when the bones of several species are laid side by side, they are more obvious to the eye than capable of being clearly described by words. As contrasted with the homologous bone of the domestic ox, the humerus of the muswa is longer, and the proximal end is thicker, but the shaft is more slender, particularly towards the elbow, and the trochleæ of that joint are deeper and more sharply defined. The median edge of the fossa magna which receives the olecranon is more elevated and acute, and the ridge which winds spirally over the upper part of the radial front of the shaft is less prominent and distinct. There are differences also in the greater and lesser tubercles, and in the notches which these prominences form. The following dimensions may be compared with those of the musk-bull and domestic cow given at p. 83.

Length from the proximal edge of the greater tubercle to the most distal point of the inner condyle of the elbow . . . . 15·7 inches.

Length from the same point to the convexity of the inner trochlea . . . . 15·6

Length from the crown of the articular ball to the distal convexity of the elbow-joint . 14·4 inches.
Circumference of the head and tubercles . . . . 14·0
Circumference of the smallest part of the shaft . . . . 5·2
Transverse diameter of distal joint . . . . 3·2
Ancono-thenal diameter of inner condyle . . . . 3·3
Ancono-thenal diameter of head and greater tubercle . . . . . . 4·7

Radius, Plate XXIV. fig. 2–2.—In length, sectional outline, and other characters, the radius of the muswa approaches nearest to that of the rein-deer, and differs so much from its homologues in the *Bovidæ* that there is no danger of mistaking it for any of them. The most striking peculiarity of form which the bone exhibits when contrasted with the radius of an ox, sheep, or of most deer, is the acuteness of the lateral or ulnar edge about the middle of its length, as represented in the section of the lower half of the bone, Fig. 2; a difference exists also in the line of junction of the radius and ulna within the elbow-joint, as exhibited in the same Plate (Fig. 2, upper figure), in the comparative smallness and acuteness of the outer triangular process of the ulna which indents into the radius.

In the position of the shaft of its *ulna*, the muswa differs from the rein-deer, and from all other ruminants whose skeletons I have seen. During a great part of its length it rests upon or coalesces with the radius at an equal distance from the median and lateral borders of this bone, as shown in section, Fig. 2, Plate XXIV. In the rein-deer the ulna coalesces with the lateral edge of the radius, there being merely a narrow groove for the lodgment of an artery to mark the limits of the two bones.

RADIUS AND ULNA.

Length of the *ulna* from the apex of the olecranon to the distal epiphysis which articulates with the cuneiform . . . . 19·8 inches.
Length of the olecranon from the horn of its semilunar trochlea to the apex of the process . . . . 4·0
Ancono-thenal diameter of the olecranon at the line of coalescence of the epiphysis . 2·9
Its transverse diameter or thickness at the same place . . . . . . 0·7
Length of the *radius* on its thenal aspect, from the proximal brim of the elbow-joint to the most distal part of the articular surface of the inner condyle . . . 16·1
Transverse diameter of the elbow-joint . . . . 2·8
Transverse diameter of the radius at its tubercle below the elbow-joint . . . 3·0
Transverse diameter of the carpal joint of the radius . . . . . . 2·7
Diameter of the radius at the coalescence of its distal epiphysis . . . . 3·0
Smallest breadth of the thenal surface of the radius . . . . . . 1·7

The extreme length of the radius figured in Dr. Buckland's Appendix to Beechey's Voyage, and which, at page 20, I have supposed to be the radius of a muswa, measures only 13·4 inches, or 2·7 less than the above, and corresponds better with the dimensions of the bone in a large buck rein-deer. It agrees also better with the latter in form, and is dissimilar to the radius of the muswa, with which I could not have confounded it had I seen the specimen. It is however two inches longer than the radius of a buck of the smaller or barren-ground rein-deer.

Carpus.—The carpal bones are as usual in the family six, four in the proximal, and two in the distal row, and they equal in size the carpus of a moderate-sized domestic ox. The *lunare* differs most in form from its bovine homologue, and the *cuneiform* is proportionally rather larger than in the *Bovidæ*. The whole height of the carpal joint is about 2¼ inches.

Metacarpals (Plate XXIV. Fig. 3).—The coalesced third and fourth metacarpals form the single cannon-bone, which is a third longer than its homologue in a full-sized ox; and though narrower in a lateral direction, has a greater ancono-thenal diameter. The anconal surface of the bone is concave or channelled, but less deeply than in the rein-deer, and the opposite side is much rounded, but marked by a fine median grove, indicating the line of coalescence of the two bones. Near the distal end of this groove there is a narrow foramen, about an inch long, corresponding to a smaller circular one in the ox. The second and fifth metacarpals are more complete than in any of the *Bovidæ*, and exceed their homologues in the rein-deer in comparative length. Of the two, the fifth is the longest, and rather the stoutest; its tapering point reaches upwards to within three inches of the carpal joint. The second is about an inch shorter. The toes connected with them have each three phalanges well developed, the ungueals being cased in small hoofs. In Plate XXIV. the second and fourth metacarpals, with their toes, are represented as occupying their natural position with regard to the cannon-bone. Their proximal acute ends do not lie close to the bone. Near the articular surface, which applies to the *unciform*, there is a small projection of the cannon-bone, which may perhaps be a coalesced rudiment of the proximal end of the fourth metacarpal.

Length of the cannon-bone from the carpal joint to the most distal curve of the ridges between the metacarpal trochleæ	13·4 inches.
Width of the proximal end of the cannon-bone	2·4
Width of the distal end	2·5
Width of the middle of the shaft	1·4
Ancono-thenal diameter of ditto on the median line	1·1
Length of the second metacarpal	7·0
Length of its three digital phalanges conjointly	3·3
Length of the fifth metacarpal	7·9
Length of its three digital phalanges conjointly	3·3
The first phalanx of the third or fourth toe measures in length . . .	3·0
The second phalanx of ditto measures	2·2
The ungueal of either measures	3·1

Sacral Extremity of the Muswa.

The *femur* is about an inch shorter than that of a full-grown domestic ox, and two inches longer than that of a young musk-bull or of an Alderney cow. It is not greatly inferior to its homologue in the camel in size. With respect to the condition of the bone, the epiphyses are scarcely so much coalescent with the rest of the bone as they are in the young musk-bull, described in page 66 and the following pages, and which has been considered to be about four years old. In all respects, but especially in the condyles, it is more slender than the femur of a full-grown domestic ox.

FEMUR.

Length from the crown of the head to the most distal edge of the rotular trochlea	17·0 inches.
Length from the crown of the head to the extreme convexity of the inner condyle	16·9
Length from the crest of the great trochanter to the extreme convexity of the outer condyle	17·7
Distance between the most elevated or proximal part of the great trochanter to the apex of the little trochanter	5·8
Greatest breadth of the trochanter	2·5

FEMUR.

Circumference of the shaft where it is most slender, or a little above the middle . .	5·1 inches.
Transverse axis of the head and trochanter	4·9
Transverse diameter of the condyles	4·1
Popliteo-rotular diameter of the knee between the inner border of the rotular trochlea and the posterior convexity of the inner condyle	5·1
Transverse diameter of the shaft where it is smallest	1·5
Popliteo-rotular diameter of the shaft at the same place	1·8

The *patella* is 2·9 inches in length, 2·1 broad, and 2·4 thick: its form being somewhat more elongated than that of its bovine homologues.

Tibia.—This bone, when compared with any of its bovine homologues, offers many distinctive characters, which are more easily gathered from a brief inspection than a lengthened description. The popliteal aspect of the shaft is not so convex as in the musk-ox, being nearly flat, and the bone presents more numerous acute lines or edges bounding flat surfaces than in the domestic ox. The following dimensions show that it greatly exceeds in length its homologue in the Alderney cow, noticed at page 86.

TIBIA.

Length from the projections in the centre of the knee-joint to the extreme point of either the inner or medial processes at the ankle	18·9 inches.
From the point to the distal edge of the fabella	19·0
Width of the head of the tibia	4·2
Width of the distal end of the bone on a line with the attachment of the fabella .	2.9
Extreme length of the heel-bone	5·3
Greatest length of the astragalus	2·9

The *tarsal* bones differ considerably in form from those of the domestic ox. The heel-bone is not only smaller but much less strong, its thickness being less than half its breadth. In the musk-ox the thickness of the astragalus is one-third greater than its breadth, and in the domestic ox the thickness a little exceeds the breadth.

The *astragalus* is also smaller than its homologue in the ox, and exhibits various minor modifications of the general form that it possesses in the ruminants.

Metatarsal.—This bone of the muswa is more slender, but above one-third longer than the corresponding bone of a full-grown domestic ox. In its shape it differs little from its homologue in the rein-deer, except that the groove on the popliteal surface is wider and shallower. The groove on the rotular surface is equally marked in both species.

METATARSAL.

Extreme length of the metatarsal	16·3 inches.
Width of its proximal end	2·1
Width of distal end above the joint	2·6
Width of shaft in the middle	2·2
The three phalanges of the hind toes measure in length conjointly *in situ* . .	8·0
The first phalanx detached is in length	3·3
The second ditto	2·4
The ungueal	3·2

CERVUS TARANDUS. *Small or Barren-ground variety.* (Tuktu of the Eskimos.)

Dr. Rae having sent me two complete skeletons, male and female, of the Barren-ground Rein-deer, killed in the vicinity of Fort Confidence, within the Arctic Circle, I subjoin the lengths of the several bones, to serve for comparison with the dimensions of larger races or species of rein-deer found in woodland districts.

	ADULT MALE.	ADULT FEMALE.
Cranium of the Rein-deer. (See page 98.)		
Length of the *skull* from the antinial tip of the premaxillary to the occipital spine	14·5 in.	12·8 in.
Length from the antinial points of the nasals along the mesial plane to the occipital spine	10·5	9·8
Lateral diameter from the edge of one squamosal to that of the other above the auditory canals	5·4	4·0
Height of the occiput from the basilar edge of the great foramen to the coronal edge of the occipital ridge	3·3	2·8
Length of the nasals at their mesial processes	4·0	4·0
Width of the face at the protuberances of the maxillaries above the fifth molars	4·5	3·8
Length of space occupied by the six maxillary molars	3·5	3·7
Length of mandible from the cutting edge of the incisors to the angle of the jaw	11·5	10·5
Antlers of the Rein-deer.		
Width of the bases of the pair of *antlers*, outside measurement	4·4	3·7
Length of the brow-process of the right antler, from its junction with the base of the antler to its most antinial snag	12·0	0·0
Basilo-coronal expansion of the brow-process of the right antler	10·5	0·0
Projection iniad in a straight line of the fourth and longest snag of the main stem of the left antler	32·0	23·8
Distance between the acute tips of the pair of antlers	21·0	5·0
Extreme spread of the antlers in an antinial-inial direction	43·0	18·0
Spinal Column of the Rein-deer.		
Length of the seven *cervicals* over their neural spines (the chord)	12·0	10·0
Ditto over their centra on the sternal surface following the curve	15·0	13·0
Proximo-distal diameter of the centrum of the *atlas*	1·3	1·2
Greatest lateral diameter of the atlas	5·1	3·9
Transverse diameter of the proximal articulating cup of the atlas	2·5	2·4
Proximo-distal diameter of the centrum of the *dentata*, excluding the odontoid process	2·7	2·3
Length of the crest of the neural spine of the dentata	3·1	2·5
Height of the distal end of the neural spine of the dentata above the arch of the neural canal	1·5	0·9
Chord of the neural spines of the fourteen *dorsals*	16·8	15·0
Length of their centra following the curve of their sternal surfaces	23·0	20·0

	ADULT MALE.	ADULT FEMALE.
Height of the neural spine of the second or of the third *dorsal* above the crown of the neural canal	6·0 in.	5·4 in.
(These are the tallest neural spines, but the tips of the fourth and fifth are at the summit of the back.)		
Length of the centra of the five *lumbars*, allowing for intervertebral spaces	9·5	8·7
Length of five *sternal pieces*, including the narrow anchylosed ensiform process	13·3	12·9
Length of the five *sacrals* on the pubal aspect (the chord)	5·2	4·9
Breadth of the first sacral proximal end	4·1	4·1
Length of seven *coccygeals*	6·3	5·5

Pelvis of the Rein-deer.

	ADULT MALE.	ADULT FEMALE.
Distance from the sternal angle of the crest of one *ilium* to that of the other	9·9	8·0
Sterno-dorsal length of crest of the ilium	3·0	2·5
Length from the sternal angle of the crest to the proximal brim of the acetabulum	6·0	5·1
Distance from the distal edge of the acetabulum to the tuberosity of the ischium	5·7	5·0
From the median edge of one acetabulum to that of the other, or breadth of the pubes	4·0	3·7
Greater axis of the foramen ovale	2·6	2·4
Distance between the rami of the ilia, measured inside	3·6	3·4

Atlantal Extremity of the Rein-deer.

	ADULT MALE.	ADULT FEMALE.
Scapula.—Length from the radial end of the glenoid cavity to the dorsal or shortest edge of the bone	10·0	8·5
Length of the dorsal edge of the *scapula*	6·0	5·6
Length of the anconal edge of ditto	9·6	8·0
Long axis of the glenoid cavity	1·6	1·4
Short axis of the glenoid cavity	1·3	1·15
The appendix of the scapula (mostly bony) is 1¼ inch broad in the female.		
Humerus.—Distance from the proximal edge of the greater tubercle to the most distal part of the lateral trochlea, being the extreme length of the bone	10·2	9·3
Length from the crown of the articular ball to the most distal articular convexity	9·6	8·8
Ancono-thenal diameter of the head and greater tubercle	2·9	2·6
Transverse diameter of the distal articulation	2·0	1·8
Ancono-thenal diameter of inner condyle	2·2	2·6
Ancono-thenal diameter of outer condyle	1·5	1·4
Radius and ulna.—Length of the *ulna* from the apex of the olecranon to the distal extremity that articulates with the cuneiform	14·1	12·7
Length of the olecranon from the horn of the semilunar trochlea to the apex of the process	2·7	2·4
Greatest width of the olecranon	1·7	1·5
Length of the *radius* on its rotular aspect from the proximal brim of the elbow-joint to the distal articular surface of the inner condyle	11·9	10·4
Smallest breadth of the rotular surface of the radius, about 2½ inches below the elbow	0·11	0·9

	ADULT MALE.	ADULT FEMALE.
Cannon-bone, or coalesced third and fourth metacarpals : its extreme length	8·1 in.	7·6 in.
Transverse diameter of the middle of the shaft of the atlantal *cannon-bone*	1·0	0·7
Width of its proximal end	1·4	1·3
Width of its distal end	1·8	1·6
Length of *second metacarpal*	2·9	2·7
Length of its three digital phalanges	2·7	—
Length of *fifth metacarpal*	3·1	2·7
Length of its three digital phalanges	2·7	—
Length of three digital phalanges of the cannon-bone	5·2	4·6

Sacral Extremities of the Rein-deer.

	ADULT MALE.	ADULT FEMALE.
Femur.—Length from the crown of the articulating ball to the distal curve of the inner condyle	11·6	10·8
Length from the crest of the trochanter to the distal convexity of the outer condyle	12·0	10·9
Transverse diameter of the ball of the femur and its trochanter	3·1	2·7
Transverse diameter of the condyles where they are thickest	2·5	2·2
Rotulo-popliteal diameter of the inner condyle and edge of rotular trochlea	3·1	2·9
Tibia.—Length from the crests in the centre of the knee-joint to the most distal point of the tibia, which articulates with the inner or mesial border of the astragalus	13·2	12·1
Transverse diameter of the head of the tibia	2·6	2·4
Width of its distal end	1·7	1·6
Length of the slender, curved, needle-shaped rudiment of the fibula attached to the outer head of the tibia*	5·6	—
Length of the fabellar, or distal end of the fibula	0·8	0·6
Length of the hinder *cannon-bone*, or coalesced *third and fourth metatarsals*	11·2	10·5
Its lateral diameter about the middle of the shaft	1·2	0·8
Its greatest rotulo-popliteal diameter, above three inches from its proximal end	1·7	1·3
Length of the three phalanges of the third or fourth hind toe	5·2	4·8
Length of the three phalanges of the second or fifth toe	2·2	2·1

As skeletons of Rein-deer are common in most of the large museums of Comparative Anatomy, it is unnecessary to add detailed descriptions of the several bones. The above measurements have been recorded chiefly to serve as guides to the size of the Barren-ground Rein-deer when contrasted with the large woodland races. Had I been possessed of authentic skeletons of the latter, I should have instituted a minute comparison between its bones and those of the Arctic animal; but this I must defer until the requisite specimens are acquired.

The antlers of the buck deer whose dimensions are recorded above in the first column of numbers, being of a good size, and exhibiting the most usual forms at the age of five years, are worth describing. The reader may at the same time refer to Plate XXIII., but he is requested to notice that the figure has been reversed in printing, the antler with the broad facial branch being in reality the right one.

The stem of both antlers has a general direction iniad for a foot and a half, and then rises

* I have not discovered this rudiment in any other ruminant.

R

coronad for two feet more, with a gradual regular curve, inclining antiniad at the tip, which is round, tapering, and acute.   From the basal ring of the right antler a branch takes a direct antiniad course, nearly parallel to the nasals, passing beyond them an inch or two, and expanding at the distance of 6 inches from its commencement into a thin plate 9 inches high in its corono-basilad diameter, and 5 inches in the other, which may be called its length.   This plate has eight marginal processes, the two next the nose being mere knobs, the succeeding ones broader and bifid, and the upper ones more pointed, longer, and acute, with their tips bent laterally.

Two inches from the facial process another branch springs from the main stem, and, projecting between laterad and coronad, expands into a plate with four prongs, which curve mesiad. A foot beyond this branch a tine about 7 inches long projects directly iniad, its tip being a little curved. This is hidden by the main stem of the antler in the point of view chosen by the artist, and consequently does not appear in the figure; it is placed exactly at the commencement of the great curvature of the stem, which is also inadequately shown in the figure.   Near the extremity of the antler there are three other tines, the lower and longest one being forked.

The facial branch of the left antler differs from the right one in being slender, round, and tapering throughout, and in curving coronad from near its commencement.   The next branch differs from its fellow of the right side merely in having but three prongs.   At the curve of the stem a tine is given off iniad, but, as on the other side, it is not represented in the figure.   There are four tines or prongs at the end of the antler.

In different individuals there is considerable variety in the number and distribution of the tines, especially with regard to the facial branch.   The fossil antlers mentioned in page 21 have a sufficiently close resemblance to those described above to leave little doubt of their having belonged to buck rein-deer.   No. 107 is the basal portion of a right antler, with a facial branch.   No. 108 is considerably decayed or water-worn, and seems to be the basal part of a left antler, having both a facial and a lateral branch.   No. 109 is a left antler broken short at the middle, the remaining part having the dimensions of the antler of the recent skeleton, but wanting the facial branch.   No. 110 is a fragment of a left antler, which retains vestiges of a slender facial branch, and of a compressed lateral one.   It is of smaller size than the antler of the recent skeleton.

The antlers of the female skeleton are much smaller than those of the male.   The right one shows merely a short subulate snag in place of the facial branch.   The succeeding branch, instead of a lateral inclination, rises coronad and antiniad, is palmated with two prongs, while its fellow of the left antler is more broadly palmated, with six prongs.   The inial branch at the bend of the main stem is much like that of the male, though smaller, and both antlers are palmated and forked at the end.

A small canine tooth is implanted in the antinial point of the maxillary in both male and female, but no similar tooth exists in the maxillaries of the muswa's cranium.

The fragment of a fossil pelvis, No. 137, p. 21, presents no appreciable difference when compared with the right acetabulum of the recent female skeleton.

The scapula, however, No. 167, p. 21, differs from its homologue in both the male and female skeletons in being somewhat more strongly built and more hollow on the internal or mesial surface. It does not agree with the scapula of either the big-horn or mountain antelope.

No. $\frac{111}{2}$, p. 21, being the middle part of the shaft of a tibia, is erroneously referred to the reindeer, and is most probably a fragment of the skeleton of a female big-horn.   If so,—and I have little doubt on the matter, after close comparisons,—another species must be added to the fauna existing at the time when the cliffs of Eschscholtz Bay were formed, and another instance of the American character of that fauna may be adduced.

No. $\frac{112}{2}$, p. 21, a fossil metacarpal bone, is a little longer than its homologue in the female skeleton, but from its slenderness it is most probably the relic of a female deer. It is identical in form with the existing species. In Plate XXII., fig. 2, the atlantal cannon-bone of the recent male deer is represented. The uppermost margin of the bone in the figure is the lateral edge with the fifth metacarpal applied to it, the second metacarpal being applied to the mesial edge of the anconal groove.

In the sacral extremities the fifth and second toes are smaller than in the atlantal ones, and their metatarsals arc wanting.

The fossil sacral cannon-bone, No. $\frac{111}{1}$, p. 21, from its relative size, is most probably a part of the same female skeleton to which No. $\frac{112}{2}$ belonged.

OVIBOS MOSCHATUS.   (Refer back to p. 66.)

It is also to Dr. Rae's kindness and exertions that I owe a fine skeleton of a musk-bull of a more mature age than that figured in Plate II., and described in the 66th and following pages. The cranium with the horns and mandible of this aged individual weigh 26¼ lbs. avoirdupois, and the bones of the whole skull are thicker and more porous than in a younger animal; the orbitar plates especially are more prominent and much rougher.

The shape of the atlas is modified a little by age. In the aged bull the neurapophysial brim of the proximal articulation is raised by an accession of bony matter, and there is also a sesamoid bone of the size of a horse-bean attached to the margin of the recent growth on each side of the vertebra. Moreover there are two small irregular ossicles at the proximal brim of the neural canal on the side of the centrum. It is probable that these sesamoid bones existed, and were even more numerous, in the young bull, but if so they were lost in cleaning the skeleton. The semilunar depression on which the accessory shoulder of the condyle plays remains deep and well defined in the old bull.

A comparison of the *dentata* of the large bull with the fossil one figured in Plate XI. (figs. 2, 3, 4), and on which the establishment of a species named *Ovibos maximus* was sought to be founded, proves that though the vertebra of the recent bull approaches the fossil in size, specific differences may be detected, and the following comparative table of dimensions yields data in confirmation of this view.

Dimensions of the dentata of the recent and fossil Ovibos.

	RECENT BULL.	FOSSIL.
Length of the centrum on its sternal aspect, from the proximal edge of the odontoid to its distal articulating surface	3·75 in.	4·2 in.
Length of centrum alone on its sternal aspect, excluding all the proximal articulation	2·65	2·8
Transverse diameter of proximal articulating surface of the centrum . .	4·4	4·6
Distance from the sternal aspect of the centrum to the nearest side of the neural canal at the atlantal end of the vertebra	2·3	2·2

	RECENT BULL.	FOSSIL.
Sterno-neural diameter of the distal articulating cup of the centrum . . .	2·4 in.	2·5 in.
Transverse diameter of the same cup	3·1	3·5
Transverse diameter of the centrum immediately above the parapophysis . .	4·0	4·3
Distance from the outside of one zygapophysis to that of the other . . .	3·2	3·7
Distance from the sternal front of the centrum at its distal end to the neural aspect of the zygapophysis	4·1	4·5
Width of the base of the neural spine at its proximal end	1·3	1·7
Length of the base of the neural spine	3·4	3·4
Thickness of the crest of the neural spine towards its atlantal end . . .	1·1	—
Greatest diameter of the articulating surface of a zygapophysis . . .	1·0	1·5

The spread of the zygapophyses, their much greater thickness at their roots, their broader articulating surfaces, and the greater thickness of the neural spine at its base, are the most obvious particulars in which the fossil differs from the recent animal.

"A Memoir on the Extinct Species of American Ox," by Dr. Joseph Leidy, was published in December, 1852, in the Smithsonian Contributions to Knowledge, which, through the kindness of the author, has just reached me. By the perusal of this very important paper, I am enabled to correct the references I have made at pages 22, 24, and elsewhere, of Dr. Dekay's *Ovibos Pallasii* to *Ovibos moschatus*. Dr. Leidy includes no fewer than twelve crania, all more or less mutilated, and all dug up in the valley of the Mississippi, in his species named *Bootherium cavifrons* (" Proceed-Acad. Nat. Scienc. vol. vi. p. 71"), to which he refers Dr. Dekay's specimen; and a comparison of his figure 2, plate iii., with the same view of the cranium of the musk-ox in plate iii. of the 'Zoology of the Voyage of the Herald,' leaves no reasonable doubt of the fossil species being quite distinct from the recent one. The accessory trochleæ of the occipital condyles are fully developed in the fossil skull, and many other peculiarities of *Ovibos* may be discerned in the figures, so that *Bootherium* and *Ovibos* are evidently very closely allied. A question now arises whether the *dentata* above referred to as the foundation of the proposed palæozoic species *Ovibos maximus*, may not be a relic of Dr. Leidy's *cavifrons*, and this might, without much risk of mistake, be decided in the affirmative, were it certain that the Siberian crania mentioned in the ' Ossemens Fossiles' (see Zool. of the Voy. of the Herald, p. 24) were identical in species with those imbedded in the drifts of the Mississippi, but as yet the evidence for such an extension of the ancient range of *cavifrons* is wanting. The size of Dr. Leidy's specimen of *cavifrons* does not exceed that of the skull of an aged musk-bull, and the *dentata* of *maximus* is of corresponding dimensions, as shown by the table given above. If the discovery of a *dentata* of the musk-ox type, and of suitable size, in the valley of the Mississippi, should hereafter fully establish the identity of *cavifrons* with *maximus*, it may be necessary to ascertain which of the two appellations was first made public. The first part of the Herald's Zoology came out in October, 1852, and Dr. Leidy's paper is dated as published in the December following. He makes reference however to the paper printed in the Proceedings of the Academy of Natural Sciences, quoted above, which I have not seen, and of which he does not record the date. Dr. Dekay's specific name of *Pallasii* is long prior to either of the above names, but until the identity of the Mississippi fossils with those found on the Ob be established, the propriety of ascribing that appellation to the American fossils remains doubtful.

Dimensions of the Skeleton of a full-grown Musk-bull.

Skull of a full-grown Musk-bull.

Length of skull from the antinial tip of the premaxillary to the occipital ridge . . 21·5 inches.
Length from the same point to the transverse suture at the root of the nasals . . 11·0
Length from the transverse suture to the occipital ridge . . . . . 10·7
Breadth at the orbits, basilar edges . . . . 10·4
Greatest breadth of the occiput behind the auditory openings . . . . 7·0
Space occupied by the molars and premolars . . . . 5·4
Greatest breadth of face just over the interval between the first and second true molars 6·2
Length of mandible from the roots of the incisors to the posterior curve above the
    angle . . . . 15·3

Spinal Column of a full-grown Musk-bull.

Chord of the cervicals measured over the neural spines . . . . 10·3
Seven cervicals measured over their centra following the curve . . . . 15·0
Chord of dorsals measured along the roots of their neural spines . . . . 24·5
Thirteen dorsals measured on the sternal aspects of their centra following the curve,
    allowing some space for intervertebral substances . . . . . 25·0
Six lumbar vertebræ measured on their sternal aspects, including intervertebral spaces . 14·0
Length of six sacrals, neural spines . . . . 9·0
Length of six coccygeals . . . . 7·5

Pelvis of a full-grown Musk-bull.

Extreme length of the pelvis . . . . 19·0
Extreme width of ditto, proximal end . . . . 14·4

Atlantal Extremities of a full-grown Musk-bull.

Length of atlantal or thenal edge of *scapula* . . . . 14·7
Length of neural edge . . . . 9·0
Length of sacral or anconal edge . . . . 13·6
Length (vertical) of scapular appendix, mostly osseous . . . . . 4·0
Length of the *humerus* from the proximal edge of the greater tubercle to the most distal
    angle of the inner condyle . . . . 13·8
Length from the crown of the articular ball to the most distal articular convexity . 12·4
Extreme length of the *radius* on its thenal aspect, measured to distal point of the inner
    condyle . . . . 12·8
Extreme length of the *ulna* . . . . 15·9
Length of the atlantal *cannon-bone* . . . . 7·6
Length of fifth metacarpal . . . . 3·2

Sacral Extremities of a full-grown Musk-bull.

Length of the *femur* from the crest of the trochanter to the most distal part of the
    outer condyle . . . . 14·9
Length from the crown of the ball to the most distal curve of the inner condyle . 14·4

Length of the *patella* . . . . 2·5 inches.
Extreme length of the *tibia* . . . . 14·7
Length of sacral cannon-bone . . . . 7·7

Dimensions of the Adult Male Aurochs Skeleton preserved in the British Museum.

This skeleton having been dismounted for the purpose of cleaning (see page 99), I availed myself of the opportunity to take the measurements of the several bones with the utmost care, and now subjoin the results for the convenience of naturalists engaged in inquiries similar to those which form the subjects of these pages.

Skull of an Adult Male Aurochs.

Length from the antinial tip of the premaxillaries to the occipital crest . . . 22·4 inches*.
Length from the antinial extremity of a nasal to the occipital crest . . . 16·8
Length of the nasals . . . . 6·8
Length from transverse suture between the face and frontal to the occipital crest . 10·1
Breadth of the face between the antinio-lateral margins of the orbits . . . 10·1
Ditto between the inio-lateral ditto . . . . 12·9
Inial-antinial or antero-posterior diameter of the orbit . . . . . 2·5
Ditto including the thickness of the brims of the orbits . . . . . 3·2
Breadth of the widest part of each nasal . . . . 2·1
Chord of the transverse breadth of both nasals at their widest part . . . 3·9
Breadth of the forehead at the lateral incurvatures between the horn-cores and orbits . 9·7
Breadth of ditto between the basal rings of the horn-cores . . . . 9·9
Breadth of ditto between the supra-orbitar foramina . . . . . . 6·6
Projection of the wall of an orbit laterally beyond the groove that runs forward from the
    supra-orbitar foramina, at the inial corner of the orbits . . . . . 3·5
Width of the face at the maxillary protuberances above the third premolars . . 7·0
Width of the antinial ends of the two premaxillaries . . . . . . 2·8
Transverse distance between their inial ends, or width of the nasal opening at that part 4·3
Length of the lateral limb of the premaxillary . . . . 5·5
Height from basal edge of the occipital foramen to upper edge of the occipital crest . 5·3
Width of occiput between the outer edges of the squamosals . . . . 9·8
Width of ditto just above the auditory canals, at a protuberance of the squamosals
    and the widest part of the occiput . . . . 10·2
Breadth of the pair of the occipital condyles . . . . 4·6
Distance between the base of the horn on its basilar aspect and the nearest part of the
    zygomatic arch . . . . 0·58
Distance from the smooth frontal process forming the neck or origin of the horn-core to
    the lateral protuberance of the occipital arch . . . . . . 2·9

* All the measurements were made by the aid of callipers, except where otherwise specially mentioned. An English inch is equivalent to 0·0254 French mètres. Some discrepancies occur between the measurements here noted and those given in page 64 and elsewhere, arising from the difficulty of obtaining the exact dimensions of the parts of a mounted skeleton. Every care has been taken to obtain correct measurements of the detached bones.

Chord of the span of the horn-cores or distance from tip to tip . . . . . 24·4 inches.

Length of one horn-core following its concave curve . . . . 9·2

Length of ditto following its convex curve . . . . 11·2

Circumference of the neck of the horn-core . . . . 9·5

Length of the space occupied by three premolars and three molars . . . 5·5

Distance between the foremost premolar and the tip of the premaxillary . . 5·8

Transverse distance between the first premolars, or breadth of the palate there . 3·2

Transverse distance between the first true molars . . . . 4·0

Transverse distance between the sixth molars . . . . 3·7

Width of the posterior nasal opening . . . . 1·8

Length of the basi-occipital . . . . 2·4

Breadth of ditto at its antinial end . . . . 1·7

Widest part of the temporal fossa behind the contraction . . . . . 1·2

Widest part of ditto before ditto . . . . 0·9

Mandible.—Distance from the roots of the incisors to the angle of the jaw . . 15·2

Distance from the roots of the incisors to the back of the condyle . . . 16·4

Distance from the posterior true molar to the angle of the jaw . . . . 4·4

Distance from the roots of the incisors to the first molar . . . . . 5·1

Length of space occupied by premolars and molars . . . . 6·2

Spinal Column of an Adult Male Aurochs.

ATLAS.

Atlas.—Length of the centrum on its sternal aspect, excluding the articulating surfaces 1·8

Length of the crown of the neural canal . . . . 2·7

Sterno-neural diameter of the vertebra, or distance between the apices of the hypapophysial knob and the neural prominence or spine . . . . . . 4·1

Sterno-neural diameter of the proximal articulating surfaces, including the slopes on the
sternal edge of the cup, against which the shoulders of the basi-occipital abut . 1·9

Sterno-neural diameter of the neural canal . . . . 1·3

Transverse diameter of the proximal articulating cup . . . . . . 4·34

Transverse width of the vertebra at the proximal ends of the lateral wings . . 5·8

Ditto at the widest part, or one-third of the length from the proximal ends . 8·05

Ditto at the sacral ends of the lateral wings . . . . 7·5

Distance between the arterial foramina on the sternal aspect, or breadth of the centrum there . . . . 3·7

Ditto on the dorsal aspect . . . . 2·4

Length of the lateral processes or wings . . . . 4·7

Length from the proximal edge of the prozygapophysis to the distal point of the lateral
wing . . . . 5·1

Transverse diameter of the distal articulating cup . . . . 4·2

DENTATA.

Dentata, Axis, or Peristropheus.—Length of the centrum on the sternal aspect, including
the odontoid process . . . . 4·0

Length of ditto on the same aspect, excluding the articulations . . . 3·0

Length of the crown of the neural canal . . . . 3·0

Sterno-neural diameter of the vertebra between the peripheral points of the hypapophysial and neural projections . . . . 6·75

Sterno-neural diameter of the proximal articulating surface . . . . . 2·7 inches.
Sterno-neural diameter of the neural canal . . . . 1·05
Height of the neural spine from the crown of the arch . . . . . . 3·4
Transverse diameter of the proximal articulating surface . . . . . 4·42
Transverse measurement, including the outsides of the zygapophyses . . . 3·4
Ditto from the apex of one diapophysis to that of the other . . . . . 5·65
Height of odontoid process . . . . 1·0
Transverse diameter of the distal articulating cup . . . . . . 2·0

Cervicals of Adult Male Aurochs.

	THIRD.	FOURTH.	FIFTH.	SIXTH.	SEVENTH.
Height or length of centrum on its sternal aspect, including the ball of the proximal articulation	3·6 in.	3·6 in.	3·7 in.	3·6 in.	2·9 in.
Ditto, ditto, excluding the ball	2·2	2·3	2·4	2·0	1·6
Length of centrum on its neural aspect	2·5	2·4	2·4	—	—
Length of the crown of the neural arch	1·6	1·65	2·0	1·5	2·1
Sterno-neural diameter of the vertebra from the hypapophysial projection to the tip of the neural spine	6·0	6·2	6·3	7·1	13·5
Sterno-neural diameter of the proximal articulating surface	2·2	2·1	2·0	—	—
Sterno-neural diameter of the neural canal	1·1	1·0	—	—	—
Transverse diameter of the proximal articulating ball	1·65	1·7	1·6	1·6	1·6
Greatest transverse distance between the outsides of the parapophyses	3·5	4·4	5·1	4·2	—
Transverse distance between the insides of the parapophyses at their ends where nearest	2·2	3·1	4·0	3·0	—
Transverse diameter of the distal articulating cup (including the sockets of ribs in seventh cervical)	2·1	1·96	2·3	2·1	2·9
Length of the vertebra from the proximal edge of the prozygapophysis to the distal point of the diapophyses	4·1	3·6	3·3	2·9	2·7
Length of the vertebra from the proximal edge of the prozygapophysis, or its metapophysis, to the distal point of the zygapophysis	3·65	3·3	3·3	3·2	3·3
Length from the proximal point of the parapophysis to the distal point of the diapophysis	4·0	4·1	3·6	3·9	—
Height of the neural spine above the crown of the neural arch	2·5	2·6	2·75	3·6	10·5
Transverse distance between the outside of one prozygapophysis and that of the other	3·6	3·7	3·8	4·0	4·2
Transverse diameter of the vertebra on the space between the prozygapophysis and the diapophysis, or at the outer walls of the arterial foramina	2·4	2·8	2·9	3·2	3·6
Transverse diameter sternad of the roots of the diapophyses and dorsad of the parapophyses	3·0	3·5	3·4	2·7	2·4
Sterno-neural diameter from the knob or brim of the					

	THIRD.	FOURTH.	FIFTH.	SIXTH.	SEVENTH.
distal articulating cup to the crown of the neural arch (the knob is broken in sixth cervical) .	3·4 in.	3·4 in.	3·5 in.	3·4 in.	3·2 in.
Distance between the outsides of the tips of the diapophyses	6·5	6·8	6·1	5·8	5·7

Dorsals (*first to fifth*) of the Adult Male Aurochs.

	FIRST.	SECOND.	THIRD.	FOURTH.	FIFTH.
Length of the centrum on the sternal aspect, including the convexity of the proximal articulating surface .	2·8 in.	2·7 in.	2·8 in.	2·8 in.	3·0 in.
Ditto, ditto, excluding the articulations . . .	2·0	2·1	2·2	2·2	2·4
Length of the centrum on the neural aspect . .	2·1	—	1·9	2·3	2·2
Length of the crown of the neural arch measured up to the root of the zygapophyses	1·8	2·2	2·4	2·3	2·4
Distance between the tip of the metapophysis (or prozygapophysis) and the brim of the distal pleurapophysial cup	2·7	—	2·7	2·7	2·5
Distance between the proximal edge of the proximal pleurapophysial cup and the distal edge of the distal one	2·4	—	—	2·3	2·5
Width of the centrum about its middle . . .	2·1	2·2	2·3	2·1	1·9
Width of the proximal articulation of the centrum, including the cups for the ribs	2·4	2·3	2·4	2·6	2·7
Sterno-neural diameter of the proximal articulation of the centrum	2·1	1·8	1·9	1·9	2·0
Sterno-neural diameter of the distal articulation of the centrum	1·4	—	2·0	2·1	2·0
Transverse diameter of the distal articulation, including the cups for the ribs	2·7	—	3·1	3·2	3·2
Atlanto-sacral axis of the base of the neural spine, including the zygapophysis	2·4	2·2	2·1	2·1	2·1
Sterno-neural diameter of the proximal end of the neural canal	1·1	0·9	0·7	0·8	1·7
Transverse diameter of the neural canal at its proximal end	1·3	1·1	1·1	1·0	1·0
Height of the neural spine above the crown of the neural arch at its proximal edge	16·7	17·0	17·2	16·2	14·8
Atlanto-sacral diameter of the extremity of the neural spine where greatest	2·8	3·5	2·3	2·3	2·2
Width of the neural spine just above the zygapophysis .	1·4	1·2	2·1	1·1	1·1
Width of the zygapophyses at their atlantal edges .	1·5	1·4	1·5	1·5	1·6

Dorsals (*sixth to tenth*) of the Adult Male Aurochs.

	SIXTH.	SEVENTH.	EIGHTH.	NINTH.	TENTH.
Length of the centrum on the sternal aspect, including the convexity of the proximal articulating surface .	3·0 in.	2·7 in.	2·7 in.	2·7 in.	2·7 in.
Ditto, ditto, excluding the articulations . . .	2·3	2·1	2·2	2·2	2·2

	SIXTH.	SEVENTH.	EIGHTH.	NINTH.	TENTH.
Length of the centrum on its neural aspect . . .	2·3 in.	2·3 in.	2·2 in.	2·2 in.	2·3 in.
Length of the crown of the neural arch, measured up to the root of the zygapophysis	2·5	2·7	2·1	2·1	2·3
Distance between the tip of the metapophysis (or prozygapophysis) and the brim of the distal pleurapophysial cup	2·5	2·7	2·7	2·9	2·6
Distance between the proximal edge of the proximal pleurapophysial cup and the distal edge of the distal one	2·5	2·5	2·3	2·5	2·4
Width of the centrum about its middle . .	1·8	1·7	1·6	1·6	1·6
Width of the proximal articulation of the centrum, including the cups for the ribs	2·5	2·3	2·2	2·1	2·0
Sterno-neural diameter of the proximal articulation of the centrum	2·0	2·1	1·8	1·8	1·7
Sterno-neural diameter of the distal articulation of the centrum	2·0	1·8	1·7	1·6	1·7
Transverse diameter of the distal articulation, including the cups for the ribs	3·0	2·9	2·6	2·4	2·5
Atlanto-sacral axis of the base of the neural spine, including the zygapophysis	2·0	1·9	1·6	1·7	1·8
Sterno-neural diameter of the proximal end of the neural canal	0·6	0·6	0·8	0·7	0·8
Transverse diameter of the neural canal at its proximal end	1·0	1·0	0·9	0·8	0·9
Height of the neural spine above the crown of the neural arch at its proximal edge	13·5	11·8	11·4	9·7	9·0
Atlanto-sacral diameter of the extremity of the neural spine where greatest	2·0	1·7	2·0	2·4	1·5
Width of the neural spine just above the zygapophysis .	1·0	1·0	0·9	0·7	0·8
Width of the zygapophyses at their atlantal edges	1·6	1·5	1·5	1·6	1·6

Dorsals (eleventh to fourteenth) of the Adult Male Aurochs.

	ELEVENTH.	TWELFTH.	THIRTEENTH.	FOURTEENTH.
Length of the centrum on the sternal aspect, including the convexity of the proximal articulation	2·6 in.	2·6 in.	2·6 in.	2·7 in.
Ditto, ditto, excluding the articulations . . .	2·1	2·1	2·2	2·1
Length of the centrum on its neural aspect . .	2·4	2·4	2·4	2·4
Length of the crown of the neural arch measured up to the root of the zygapophysis	—	2·0	2·2	2·3
Distance between the tip of the metapophysis (or prozygapophysis) and the brim of the distal pleurapophysial cup .	2·4	2·7	3·2	—
Distance between the proximal edge of the proximal pleurapophysial cup and the distal edge of the distal one .	2·3	2·5	2·5	—
Width of the centrum about its middle .	1·6	1·6	1·8	2·0

	ELEVENTH.	TWELFTH.	THIR-TEENTH.	FOUR-TEENTH.
Width of the proximal articulation of the centrum, including the cups for the ribs	2·0 in.	2·0 in.	2·2 in.	2·6 in.
Sterno-neural diameter of the proximal articulation of the centrum	1·7	1·7	1·7	1·7
Sterno-neural diameter of the distal articulation of the centrum	1·7	1·7	1·6	1·7
Transverse diameter of the distal articulation of the centrum, including the cups for the ribs	2·4	2·5	3·0	2·2
Atlanto-sacral axis of the base of the neural spine, including the zygapophysis	1·7	1·7	2·6	3·0
Sterno-neural diameter of the proximal end of the neural canal	0·8	0·7		0·8
Transverse diameter of the neural canal at its proximal end	0·8	1·0	0·9	1·0
Height of the neural spine above the crown of the neural arch at its proximal edge	7·5	6·2	5·0	4·2
Atlanto-sacral diameter of the extremity of the neural spine where greatest	1·3	1·0	1·3	1·7
Width (lateral thickness) of the neural spine just above the zygapophysis	0·8	0·9	0·8	1·1
Width of the zygapophyses at their atlantal edges	1·7	1·2	1·5	1·7

Lumbars of an Adult Male Aurochs.

	FIRST.	SECOND.	THIRD.	FOURTH.	FIFTH.
Length of the centrum, including the proximal articulation (on the sternal aspect)	2·6 in.	3·0 in.	— in.	2·9 in.	2·7 in.
Length of the centrum, excluding the articulation	2·2	2·2	2·3	2·3	2·2
Length of the centrum on its neural aspect	2·5	2·7	2·5	—	2·3
Length of the crown of the neural canal up to the base of the zygapophysis	2·3	2·5	2·4	2·5	2·0
Distance between the apex of the metapophysis and the distal end of the zygapophysis	3·9	4·1	4·0	3·8	4·1
Width of the centrum at the root of the diapleurapophysis	2·1	1·9	2·0	2·2	2·7
Transverse diameter of the proximal articulation of the centrum	2·0	1·9	1·9	2·1	2·2
Sterno-neural diameter of ditto	1·8	1·9	1·9	1·9	1·7
Transverse diameter of distal articulation	2·2	2·2	2·3	2·6	3·2
Sterno-neural diameter of distal articulation	1·8	1·9	1·8	—	1·4
Transverse projection of a single diapleurapophysis	5·0	6·1	6·5	6·8	5·4
Distance from the tip of one pleurapophysis to that of the other	11·7	13·7	14·7	15·0	11·7
Distance between the outside of one prozygapophysis and that of the other	2·6	2·9	3·1	3·2	3·6
Distance between the outside of one zygapophysis and that of the other	2·0	2·0	2·0	2·3	2·5
Height of the neural spine above the crown of the neural arch	4·1	4·0	3·6	3·3	3·0
Sterno-neural diameter of the neural canal	0·9	0·8	0·9	0·7	1·0
Transverse diameter of the neural canal	1·0	1·1	1·0	1·4	1·2

Sacrum of an Adult Male Aurochs.

Transverse diameter of the proximal end of the sacrum between the edge of one lateral
    process and that of the other .   .   .   .   8·5 inches.

Transverse diameter of the proximal articulation .   .   .   .   .   3·2

Sterno-neural diameter of the proximal articulation .   .   .   .   1·4

Transverse distance between the insides of the prozygapophyses .   .   .   2·5

Transverse distance between the outer shoulder of one prozygapophysis and that of the
    other .   .   .   .   4·4

Width of the neural canal at its atlantal end .   .   .   .   .   1·8

Sterno-neural diameter of the neural canal at its atlantal end .   .   .   0·9

Distance between the proximal neural foramina on the sternal aspect .   .   .   1·9

Distance from the proximal edge of the prozygapophysis to the distal point of the sixth
    sacral .   .   .   .   12·5

Greatest width of second sacral (at its junction with the first) .   .   .   5·5

Greatest width of third sacral (at its junction with the second) .   .   .   4·3

Greatest width of fourth sacral (in its middle) .   .   .   .   .   3·4

Greatest width of fifth sacral (in its middle) :   .   .   .   .   3·5

Greatest width of sixth sacral (at its distal extremity) .   .   .   .   3·6

Length of the centrum of first sacral on its sternal aspect .   .   .   .   2·1

Pelvis of Adult Male Aurochs.

Distance between the crest of the ilium and the tuberosity of the ischium .   .   22·5 inches.

Distance between the crest of the ilium and the atlantal edge of the acetabulum .   12·2

Distance between the atlantal edge of the acetabulum and the tuberosity of the ischium 12·5

Distance between the crest of the ilium and the atlantal end of the ischiatic notch .   9·2

Length of the neural opening of the ischiatic notch .   .   .   .   10·0

Distance between the conical protuberance of the ischium and the apex of the neural
    aspect of that bone .   .   .   .   4·1

Length of the foramen ovale .   .   .   .   4·5

Vertical diameter of the foramen ovale .   .   .   .   .   .   2·6

Length of the branch of the pubal, extending from the symphysis to the brim of the
    acetabulum .   .   .   .   4·2

Chord of the ilium where widest (near its crest) .   .   .   .   10·0

Sterno-neural diameter of the stem of the ilium at its narrowest part .   .   2·3

Sterno-neural diameter of the pubal, proximal branch .   .   .   2·6

Chord of the distal end of the ischium .   .   .   .   .   .   5·8

Length of the symphysis of the pubis down to the angle formed by the meeting of the
    two ischia .   .   .   .   8·7

*The Ribs, measured from the tubercle on their shoulders to their sternal extremities, have chords of the
following lengths :—*

First, 12 inches; second, 13·8 inches; third, 15·2 inches; fourth, 16·9 inches; fifth, 17·6 inches; sixth,
19·0 inches; seventh, 20·1 inches; eighth, 20·7 inches; ninth, 21·3 inches; tenth, 20·9 inches; eleventh,
20·5 inches; twelfth, 18·8 inches; thirteenth, 17·0 inches; fourteenth, 14·2 inches.

Scapula of an Adult Male Aurochs.

Length of its distal edge	16·4 inches.
Length of its thenal edge	18·8
Length of its neural edge (or base)	11·6
Length of the mesial line measured on the interior or mesial aspect	18·5

Humerus of an Adult Male Aurochs.

Length from the crest of the greater tubercle to the most distal point of the outer trochlea	15·5 inches.
Length from the crest of the lesser tubercle to the most distal curve of the inner condyle	14·5
Length from the crown of the ball to the middle anconal trochlea on the mesial plane	13·5
Radio-ulnar diameter of the proximal end of the humerus	5·1
Ancono-thenal diameter from the outside of the greater tubercle to the opposite convexity of the ball	5·7
Ancono-thenal diameter of the middle of the shaft	2·5
Circumference of the shaft in its middle	7·5
Ancono-thenal diameter of the outer condyle	3·7
Ancono-thenal diameter of the inner condyle	2·7
Greatest transverse diameter of the condyles	4·2
Greatest breadth of the distal articular surface	3·6

Radius and Ulna of an Adult Male Aurochs.

Distance from the anconal edge of the proximal end of the *radius* to the most distal point on the inner side	13·4 inches.
Distance from the anconal edge of the proximal end of the radius to the most distal point on the outer side	13·2
Transverse diameter of the articulating surface at the elbow-joint	3·6
Transverse diameter of the shaft near its middle	2·1
Transverse diameter of the bone at the tubercles above the carpus	3·5
Transverse diameter of the surface that articulates with the carpals	2·1
Ancono-thenal diameter of the middle of the shaft	1·3
Extreme length of the *ulna*	18·0
Ancono-thenal diameter of the olecranon near the joint	2·3
Distance from the extremity of the olecranon to the radial brim of the elbow-joint	5·9
Ancono-thenal diameter of the olecranon close to the elbow-joint, below the crescentic beak	2·3

Metacarpus of an Adult Male Aurochs.

Extreme length of the *cannon-bone*	7·8 inches.
Width of its distal articulation	2·9
Width of the middle of its shaft	1·9
Width just below the carpal joint	3·1
Ancono-thenal diameter of the middle of the shaft	1·2
Length of the second metacarpal	1·8
Length of the fifth metacarpal	2·4

Femur of an Adult Male Aurochs.

Length from the convexity of the head of the bone to the most distal part of the inner
    trochlea . . . . 17·5 inches.
Length from the crest of the greater trochanter to the outer edge of the outer trochlea 18·8
Length from the crest of the greater trochanter to the outer curve of the outer condyle 18·1
Length from convexity of the head to the convexity of the inner condyle . . . 17·3
Transverse diameter of the trochanter and head of the femur . . . . 5·7
Distance measured from the rotular corner of the crest of the trochanter to the apex of
    the lesser trochanter . . . . 6·4
Transverse diameter at the condyles . . . . 4·5
Transverse diameter of the middle of the shaft . . . . . 1·8
Rotulo-popliteal diameter of the middle of the shaft . . . . . 2·1
Rotulo-popliteal diameter of the inner condyle . . . . 5·8
Rotulo-popliteal diameter of the outer condyle . . . . 6·1

Tibia of an Adult Male Aurochs.

Length of the bone on its rotular aspect and mesian line . . . . 17·6 inches.
Length from the crests for the crucial ligaments to the extremity of the outer condyle . 18·8
Transverse diameter of the knee . . . . 4·8
Rotulo-popliteal diameter of the head of the bone measured from the spine of the tibia
    to the popliteal margin of the joint . . . . . . 4·1
Transverse diameter of the distal end of the tibia . . . . . . 2·7
Rotulo-popliteal diameter of the shaft a little below its middle . . . 1·8
Lateral diameter of the shaft a little below its middle . . . . . 2·1
Length of the tarsals *in situ*, excluding their indentations into the end of the tibia . 2·5

Calcaneum of an Adult Male Aurochs.

Length of its outer side . . . . 6·4 inches.
Rotulo-popliteal diameter of its shaft . . . . . . 1·8
Circumference of its shaft . . . . 5·2

Sacral Metatarsal of an Adult Male Aurochs.

Length of the outer side of the cannon-bone . . . . . 9·8 inches.
Length of the inner side of the cannon-bone . . . . . 9·5
Lateral diameter of its proximal end . . . . . . 2·4
Lateral diameter above the distal joint . . . . . . 2·6
Lateral diameter in the middle of its shaft . . . . . 1·6
Rotulo-popliteal diameter of its head . . . . . 1·4

Phalanges of the Sacral Extremities.

Length of the *first phalanx* . . . . 3·0 inches.
Width of its proximal end . . . . 1·5
Width of its distal end . . . . 1·4
Length of the *second phalanx* . . . . . . 2·1
Width of its proximal end . . . . 1·4

Width of its distal end   .   .   .   . 1·3 inches.
Length of the *ungueal* on its rotular side   .   .   .   . 2·8
Length of the *ungueal* on its popliteal side .   .   .   . 3·3

ROCKY MOUNTAIN ANTELOPE.

(*Rupicapra* and *Antilope Americana*, Blainville; *Antilope lanigera*, and *Aplocerus*, Ham. Smith;
Capra Americana, Richardson, Fauna Amer.)

(Osteology, Plates XVI., XVII., XVIII., XIX.)

Cranium.   (Plates XVI., XVII.)

The general contour of the coronal aspect of the skull partakes much of the ovine type, but the face is more lengthened.   A perfectly straight profile is prolonged throughout the nasal bones, and in the frontal to near the supra-orbital foramina, beyond which the forehead arches gently, and gradually attains its summit between the horns (Plate XVI.).   From thence the profile descends with some rapidity to the occipital ridge, showing in its progress some gibbosity at the union of the parietals and interparietals.   An angle of about 135 degrees is formed by the meeting of the facial and parietal plates; and the occipital plane, which lies beneath the ridge, joins the general plane of the parietals and super-occipitals at an angle of about 100 degrees.   The facial line and occipital one, if produced, would meet in an acute angle.   In our specimen the sagittal suture remains perfect between the frontals, but has disappeared between the parietals.

Transversely the forehead is somewhat hollow, owing to a slight prominence of the orbital sockets (Plate XVII.).   The horn-cases rise neatly from the corono-inial aspect of these processes, stand at the the distance of one of their own basal diameters apart from one another, and at rather less than a diameter from the nearest edge of the orbit.   In section the horn-cases are oval, or very nearly round, the lateral diameter being only slightly less than the fronto-occipital one.   In form they are conical, and in their coronad rise they incline iniad, without perceptible curvature, though the inial side is rather shorter or less full than the antinial one.

At a little distance behind the bases of the cores the coronal suture takes a directly transverse course, then curves forward a little, prior to descending laterally behind the orbital projections. This curvature is small in comparison with that which the same suture takes in the domestic sheep.   The eye-sockets are not so prominent, and their walls are thinner than in the sheep, but the parietals are shaped much as in that species, and the interparietals coalesce with them at an equally early age. On their lateral aspects the conjoined bones are more convex than in the sheep, and a better-defined but only slightly prominent ridge is developed between the lateral and coronal faces of each interparietal.   In the occipital suture and general form of the occipitals this antelope much resembles the common sheep; but the exoccipital plane meets the superoccipital one more acutely, and the ridge is consequently more sharply defined.   The two articulating surfaces of each condyle meet rather more acutely, and are less convex than in the sheep.   When viewed on their basilar aspect the condyles are seen to project more than in the last-named animal, and to be separated from one another by a deeper notch.

The basi-occipital and basi-sphenoid are narrower and more concave than in the sheep, the concavity being produced by the prominence of the shoulders of the former bone, and of the lateral tubercles which exist at the coalescent ends of the two bones ; but the mesial or hypapophysial ridge of the basi-sphenoid is less evident than in the sheep. In the paroccipitals, and exoccipital spines, there is a general resemblance to the homologous bones of the sheep ; and the same general likeness extends to the ali-sphenoids, the posterior nasal openings, the palatines, and the palatine plates of the maxillaries.

On their lateral aspects the maxillaries appear longer and less deep than their homologues in the sheep, and they emit a more prolonged and acute anterior process, which requires a more oblique suture with the premaxillary.

A more rectangular form is possessed by the lachrymal than in the common sheep, and the bone has a greater width, chiefly at the expense of the malar, which again differs considerably in shape from the malar of the sheep, and is connected with the maxillary by a deeply angularly indented suture. The premaxillaries, more oblique and longer than in the sheep, have also more dilated semicircular antinial extremities, while their inial ends are pointed, and do not reach so far back as the junction of the nasals and the maxillaries.

Teeth.

Incisors, $\frac{0-0}{3-3}$; Canines, $\frac{0-0}{1-1}$; Premolars, $\frac{3-3}{3-3}$; Molars, $\frac{3-3}{3-3}$; Total, $\frac{6-6}{10-10}$.

In the general form of the teeth the resemblance is pretty close both to the common sheep and to the prong-horned antelope, but the incisors differ in being all nearly of the same size, and the incisiform canine is but little narrower. There is no trace of a tooth in the antinial process of the maxillary, and the upper molar series stand in slightly curved lines converging a little anteriorly. The three premolars have one-lobed crowns, and their flattish lateral surfaces are bounded before and behind by rounded pillar-like folds. All the molars are bilobed, with, on the exterior surface, a central pillar-like roll and two marginal ones, as prominent as in the sheep. On the mesial surface the divisions of the teeth are rounded, as usual, but are destitute of the short intermediate fold which exists in the molars of an ox. The mandibular teeth are not implanted in so curved a line as the maxillary molars, and the first premolar is smaller and more cylindrical than its opponent of the upper series. A small and not very distinct posterior lobe exists in the third mandibular premolar, and the third true inferior molar has also a comparatively small and inclined third lobe. No vestige of a third lobe can be perceived in the opposing tooth of the maxillary series. The pillar-like folds on the mesial sides of the mandibular molars are not well defined, and the central one of each tooth is wanting, except on the third molar, which has a narrow pillar between the second and third lobe.

Measurements of the Cranium with callipers.

Length of the skull from the antinial end of the premaxillary to the occipital ridge   . 11·1 inches.
Distance from the same point to the intersection of the transverse nasal suture with the
    sagittal suture   .   .   .   . 5·3
Breadth of skull at the coronal borders of the orbits   .   .   .   .   . 2·8
Breadth of skull at the basal borders of the orbits   .   .   .   .   .   . 4·4
Breadth of the parietals behind the orbital processes of the frontal   .   .   . 2·8
Breadth of the skull at the everted edges of the occipitals behind the auditory tubes   . 2·9

Distance from the lateral edge of one condyle to that of the other . . . . 3·1 inches.

Distance from the occipital suture to the basilar margin of the occipital ridge . . 1·1

Distance from the occipital suture to the basilar edge of the foramen magnum, mesial line 2·5

Distance from the occipital ridge to the proximal edge of the foramen magnum . . 1·3

Distance from same point to the basilar edge of same foramen . . . . . 1·9

Distance from basilar edge of foramen magnum to antinial end of the basi-sphenoid . 2·2

Breadth of the shoulders of the basi-occipital . . . . 1·2

Breadth of the basi-sphenoid at its antinial end . . . . 0·4

Greatest width of the posterior nasal opening . . . . 0·6

Chord of the space occupied by the series of six maxillary molars and premolars . . 3·1

Distance between the foremost premolar and the antinial tip of the premaxillary . . 3·1

Breadth of the palate between the foremost premolars . . . . 1·2

Breadth of the palate between the first pair of true molars . . . . . 1·5

Breadth of the palate between the second lobes of the last pair . . . . . 1·6

Distance between the third lobe of last mandibular molar to the posterior curve of the angle of the mandible . . . . 2·3

Distance from the roots of the mandibular incisors to the posterior border of the condyle 8·6

Distance from the roots of the mandibular incisors to the posterior curve above the angle of the mandible . . . . 8·3

Rise of the coronoid process above the surface of the condyle . . . . . 1·4

Breadth of the face from the outer side of one maxillary at the fangs of the first true molar to the same place of the other maxillary . . . . 3·3

Vertebral Column.

Cervicals.—The chord of the *cervicals* measured over the neural spines is $7\frac{1}{2}$ inches, and the curve of the sternal surface of their centra measures $9\frac{1}{2}$ inches. (Plate XVIII.)

Atlas.—A deep articular cup for the reception of the occipital condyle is formed as usual by the neurapophyses or prozygapophyses and the raised shoulders of the centrum, which are rounded in outline, and the lateral notch between the two processes is wide and shallow. The distal articulating surface of the atlas is oblique and slightly concave transversely, but is decidedly convex in its sterno-neural diameter, and as the corresponding surface of the dentata is convex in both directions, there must be considerable play between them. When the two vertebræ, however, are viewed on their sternal aspect, the line of articulation appears as a segment of a large curve, convex atlantad, by which some firmness is given to the joint in its lateral motions. A mere tubercular knob stands in place of a neural spine, and a small mesial ridge runs from it to the proximal margin of the neural canal. A hypapophysial spine is more defined, occupying the distal mesial point of the centrum, and projecting a little sacrad of the adjacent articulating surface. Security is added to the joint by the distal corners of the transverse processes also drooping sacrad. On the whole, the atlas of this antelope departs but little from the ovine type.

Dentata.—Plate XVIII. fig. 1 gives the profile of the second cervical, and shows the arched peripheral outline of its neural spine. Figure 2 of the same plate represents its sternal aspect and the corset shape, which may be considered as the typical form of this bone in the ruminants. Its hypapophysial ridge is prominent but not acute, and its distal extremity stands out sacrad, rendering the postarticulating surface oblique. There is a degree of obliquity in the antarticulating surface

T

also, combined with some convexity, so that this joint makes a nearer approach to the same part in the horse than in the ox.

The *third, fourth,* and *fifth cervicals* differ less in general form from one another than the homologous vertebræ of the ox do, but they may be distinguished among themselves by the transverse processes of each succeeding one terminating more remotely from the distal articulating surface of the centrum, and showing more distinctly its composition of diapophysis and parapophysis, and by the lengthening of the neural spines. The hypapophysial ridge is most acute in the fourth cervical, and is nearly obsolete in the fifth one, while the sixth has no vestige of a ridge whatever, except a slight protuberance at its distal end. In this last-mentioned cervical the peripheral edges of the parapophyses are longer than their bases, and their pleurapophysial borders are narrow, and, in our specimen, but partially coalescent. The extent of this border is shown in Plate XVIII. figure 1. The diapophysis also represented in the same figure is very short. A very slight indication of a hypapophysial ridge exists in the seventh cervical, also at its distal end. It is more strongly marked in fig. 2, Plate XVIII., than in the bone itself. In this vertebra the diapophysis is more directly lateral and stouter than in the other cervicals, and the neural spine is very considerably larger in all its dimensions than the spines of the four immediately preceding ones, in which that process is comparatively small.

Dorsals.—The thirteen dorsal vertebræ measure over their neural spines $13\frac{1}{4}$ inches, and along the sternal faces of their centra $14\frac{1}{2}$ or $14\frac{3}{4}$ inches, some allowance being made for the shrinking of the intervertebral substances in drying. Of the neural spines the third is the longest absolutely, and measures from the edge of the socket for the shoulder of the rib 4·2 inches; but the spines of the second and fourth are scarcely a line shorter, and the tips of the fourth and fifth form the apex of the dorsal curve in the skeleton. The other preceding and following spines diminish very gradually in length, the arch of the back being small. None of the dorsal spines recline strongly sacrad, and the tenth, which is nearly vertical, is not symmetrical, its anterior edge having the reclination of the spines which precede it, and the posterior edge the vertical direction of the lumbars. In form the eleventh, twelfth, and thirteenth spines resemble those of the lumbars. In the twelfth and thirteenth the articulating surfaces of the prozygapophyses face mesiad, as in the lumbars, so as to receive the zygapophyses of the preceding vertebræ within them. Moreover, the twelfth dorsal has a facet on its diapophysis for articulation with the rib, which occupies the whole tip of the process, without any point projecting sacrad. This facet is wanting in the thirteenth, whose diapophysis is larger and more conical, and there is no tubercle to the corresponding rib.

A distinct metapophysis exists on the back of the diapophysis of the second dorsal. This process is most elevated and acute in the fourth dorsal, but is nearly as large in the succeeding ones, and it recedes mesiad of the articulating facet, and projects more atlantad in each succeeding vertebra. On the twelfth and thirteenth it has changed its form and situation so far as to become a mere rounded protuberance on the back of the prozygapophysis, a change rendered necessary by the position of the oblique articulations of these dorsals.

The dorsal centra lengthen successively from the first to the thirteenth, and their sternal faces become less rounded and more ridged, but none of them are so much pinched in at the sides as in the homologous bones of the *Bovinæ*, and only the last three or four in the series are so acute as to merit the appellation of keeled.

The Ribs

are thirteen in number, of which the ninth and tenth are the longest when measured from their respective shoulders, the first being the shortest, then the second, and thirdly the thirteenth. None of them are much dilated in their course, and the fourth, which is the widest in the blade, is very little broader than the preceding or three following ones. The first rib is altogether the stoutest, and also the widest at its union with the sternal rib or cartilage.

Sternum.

Seven bony pieces and the ensiform cartilage, which is almost wholly ossified by the deposition of granular concretions, compose the sternum. In the aggregate these bones form an arc, most curved at its atlantal end, but the *manubrium* does not make an angle with the succeeding pieces. The fourth, fifth, and sixth segments are the broadest, the others being considerably narrower. In the insertion of the cartilages or sternal ribs the usual ruminant type is adhered to. The first pair of cartilages embrace the apex of the manubrium; the succeeding ones, down to the sixth pair, are articulated to sockets formed in the edges of two adjoining sternal pieces; the seventh pair has its sockets sunk in the side of the seventh sternal piece only, and the sockets of the eighth pair have their sockets common to the seventh and eighth sternal segments. The remaining cartilages do not reach the sternum, but the ninth pair are in part coalescent with the eighth. All are rigid, from the quantity of osseous matter by which most of the cartilage has been replaced. The sterno-neural internal diameter of the thorax in the skeleton at the seventh sternal piece is in the attitude of inspiration nearly 11 inches.

The Lumbars

are six in number. Their centra are little compressed or excavated on the sides, but all of them have a keel more or less distinctly indicated, though it is not prominent in any of them. None of them exceed the last dorsal in length, and the sixth is shorter and wider. All their pleurapophyses are inclined forwards. On the third, fourth, and fifth the metapophysis is rather acute, and projects a little atlantad; this process is scarcely so conspicuous on the sixth, and it is obsolete on the first and second.

Sacrals and Caudals.

The five *sacrals* measure along their neural spines 5 inches, and the chord of the arc they form on their pubal face is a quarter of an inch longer. The last three are nearly of one width, and are rather less than half the breadth of the first one. The nine *caudals* are 6 inches long.

Pelvis.

This basin has at its atlantal brim a sacro-pubal diameter of 5½ inches, and a transverse one between the rami of the ilia of 3½ inches. These rami are long and parallel, and the brim is completed by the pubes and sacrum curved in arcs of a circle. Little expansion of the *ilium* takes place towards its sternal corner, into which the bone tapers suddenly and acutely. The crest of the bone is thickened as usual, and the ridge which descends through the dorsum towards the acetabulum is much rounded and little elevated.   No other notch exists in the brim of the acetabulum than the one which is contiguous to the atlantal edge of the *foramen ovale*.   This last-named aperture is obtusely oval, with a long diameter of 2 inches and a transverse one of 1·3 inch.   The symphysis of the pubes and the distal margin of the ischium is composed of one three-branched piece, fully ossified, but not in our specimen coalescent with that bone and the pubal.   The ilium, pubal, and ischium, however, have entirely coalesced with one another, so that no trace of their lines of junction are discernible even in the acetabulum.

Length from the crest of the ilium to the tuberosity of the ischium   .   .   .   . 10·0 inches.
Distance from anterior angle of the crest of one ilium to the same point of the other   .   6·8
Mean diameter of the brim of the acetabulum   .   .   .   . 1·2
Distance between the edges of the rami of the ischium on the neural aspect of the pelvis 1·5

Atlantal Extremities.

In form the scapula scarcely presents a tangible point of difference from its homologue in the big-horn (page 94), except that the distal or humeral end of the spine, instead of being straight, is concave.   Their glenoid cavities differ slightly from one another in those of the antelope, having a broadly oval brim without a marked notch.

SCAPULA.
Length of the thenal or coracoid edge of the scapula, excluding cartilage   .   .   . 7·5 inches.
Length of its anconal edge   .   .   .   . 7·2
Length of its dorsal edge or base .   .   .   . 4·8
Lateral diameter of the glenoid cavity .   .   .   . 1·2
Theno-ulnar diameter of ditto   .   .   .   . 1·3

Though this animal has much shorter limbs and a less stately form than the big-horn, its humerus is comparatively longer, as may be perceived by contrasting its dimensions with those recorded in page 94.

HUMERUS.
Length from the crown of the ball to the extremity of the inner condyle   .   .   . 7·8 inches.
Length from the tip of the greater tubercle to the extremity of the outer condyle .   . 8·4
Diameter of the head and greater tubercle   .   .   .   . 2·5
Transverse diameter of the condyles   .   .   .   . 1·8

Some slight differences exist in the form of the greater tubercle of the humerus of this antelope and that of the big-horn, and the thenal ridge in the upper part of the shaft of the former is rougher and better defined.

In the length of its *radius* and *ulna* the mountain antelope falls much short of the big-horn. These bones are moreover less arched thenad, and the *radius* has obtuse lateral edges, while the inner edge of that bone in the big-horn is almost as acute as in the moose-deer.   Still more decidedly does the *ulna* differ from its homologue in the big-horn in not being appressed to the radius and adherent to it by a broad surface, but compressed, thin, and coalescent with the radius by a comparatively very narrow edge.   It is thin and more elevated from the surface of the radius in its middle than in the common goat, moose-deer, or domestic ox (see Plate XIX. fig. 2).   The line of junction of the two bones within the elbow-joint much resembles the same part in the big-horn, except that the lateral process of the ulna is acutely triangular, and not rounded at the tip as in the big-horn, and the mesial projection of the radius is truncated, leaving exposed an irregular pit and furrow in the ulna for the reception of a fatty synovial gland. An inspection of Plate XIX. fig. 1 will give a better idea of the form of this joint than a verbal description. In the big-horn the small descending lateral process of the ulna within the elbow, besides being rounded at its tip, has a wider exterior shoulder. In the elbow of the musk-bull, figured in Plate XV. fig. 5, the angular process is much less acute, and the shoulder is wanting, as it is in the *Bos crassicornis* (fig. 4 of the same plate).

	MOUNTAIN ANTELOPE.
Length of the *radius* on its thenal aspect and mesial line	7·5
Width of the radius at its proximal end	1·7
Width of the radius at the middle of its shaft	0·95
Theno-anconal diameter of its middle	0·55
Length of the *ulna* from the apex of the olecranon to the distal extremity of the bone	9·9
Length of the olecranon above the horn of its crescent, which forms the upper edge of the elbow-joint	2·2

Carpus.

This joint is about an inch in height, and consists as usual of six bones in two rows, four belonging to the proximal row and two to the distal one. In their forms the carpal bones resemble those of the big-horn.

In the shortness of the *metacarpal* the mountain antelope presents a strong contrast to the long slender limb of the big-horn. It is even proportionally shorter than in the domestic sheep, as the following dimensions will prove on comparison with the measurements noted in page 95.

	MOUNTAIN ANTELOPE.
Length of the metacarpal	4·2 inches.
Its width at the carpal joint	1·35
Its width in the middle	1·1
Its width above the distal joint	1·7
Ancono-thenal diameter near the middle	0·52

A slender fusiform rudiment of a fifth metacarpal, rather exceeding half an inch in length, is articulated to the cannon-bone about two lines below the carpal joint. No vestige of a second metacarpal was detected. The small pair of hoofs attached to the distal end of the metacarpal contain each a small compressed bone, representing apparently the ungueal phalanx of the second and fourth toes.

As usual, all the phalanges of the fore toes are a little stouter than the hind ones.

Length of the first phalanx of fore toe	1·7 inches.
Length of the second phalanx of a fore toe	1·2
Length of the ungueal of a fore toe	1·7

Sacral Extremities.

A smaller curvature rotulad, and a greater prominence of the lesser trochanter, are the principal differences of form that catch the eye on comparing the *femur* of this antelope with that of the big-horn.

Length from the crown of the head of the femur to the distal convexity of the inner
 condyle . . . .   9·2 inches.
Length from the crown of the head of the femur to the distal curve of the trochlea .   9·4
Length from the tip of the trochanter to the convexity of the outer condyle . .   9·3
Diameter of the trochanter and articulating ball . . . . . .   2·45
Transverse diameter of the two condyles . . . .   2·1
Rotulo-popliteal diameter of the inner ridge of the trochlea and inner condyle .   2·5
Rotulo-popliteal diameter of the outer ridge of the trochlea and outer condyle .   2·2
Circumference of the shaft in its middle . . . .   3·0
Its lateral diameter there . . . .   0·84
Its rotulo-popliteal diameter there . . . .   0·9

From these measurements, when compared with those of the big-horn recorded at page 96, we may perceive that the femur of the antelope has not quite so much relative length as the humerus. Some differences will also be discovered in the developments of the condyles in the respective species.

The knee-pan is more lengthened in the Rocky Mountain antelope than in the big-horn, and the *tibia* is much less arched poplitead than in either the musk-ox or big-horn.

Length of the *tibia* from the crucial crests to the tip of the outer malleolar process .   9·7 inches.
Length from the crucial crests to the tip of the inner malleolar process . . .   10·2
Width of the head of the tibia . . . .   2·1
Width of the tibia at its distal end . . . .   1·4
Length of the *fabellar* . . . .   0·6
Breadth of ditto . . . .   0·9

No other trace was discovered of a fibula than the irregular bone named *fabellar* by Professor Owen.

Length of the *calcaneum* . . . .   3·0 inches.
Length of the *metatarsal* . . . .   4·6
Width of the *metatarsal* just above the metatarsal joint . . . .   1·5
Width of the *metatarsal* one-third of its length from the joint . . . .   0·8

The tarsal joint consists, as is usual in the ruminants, of four bones besides the fabellar and two sesamoids. The *metatarsal* (see Plate XIX. fig. 5, 6) is greatly shorter than that of the big-horn, and resembles in its dimensions the homologous bone of the domestic sheep. On its popliteal aspect close to its proximal end and nearly on the median line there is a thin, orbicular, sesamoid bone, a quarter of an inch in diameter. A similar bone exists in the same situation in the big-horn, but I have detected none in the muswa, rein-deer, or ox.

Length of the first phalanx of the hind toe . . . . . .   1·7 inches.
Length of the second phalanx of the hind toe . . . . . .   1·3
Length of the ungueal phalanx of the hind toe . . . . . .   1·6

An ungueal bone exists in each of the accessory hoofs, representing the second and fifth toe. The sesamoids are of the usual number, viz. four at the distal trochleæ of the metatarsal and one at the distal end of each second phalanx.

In closing these osteological notices, which have been drawn up for the purpose of elucidating the North American fossil fauna of the drift and more recent epochs, we may recur to Dr. Leidy's "Memoir on the Extinct Species of American Ox." Of these he enumerates five, ranged below in a column parallel to the probably or possibly synonymous appellations of the species, described in the Zoology of the Voyage of the Herald.

LEIDY.	RICHARDSON.
Big-bone Lick and vicinity.	*Eschscholtz Bay.*
Bison Americanus?	Bison priscus?
Bison latifrons.	
Bison antiquus?	Bison crassicornis.
Bootherium bombifrons.	
Bootherium cavifrons	Ovibos maximus.
	Ovibos moschatus.

Of the fossil bones referred by Dr. Leidy to *Bison Americanus,* the crania only were compared with their homologues in the existing species, and their specific identity he himself states cannot be positively affirmed. It is probable that the resemblance is no greater than that which prevails among the bones which I have described in the preceding pages under the name of *Bison priscus?* (p. 53.) These I found to resemble more closely the corresponding parts of *B. Americanus* than of *B. Europæus,* but to differ decidedly from both. I did not possess the means of making the requisite comparisons with the fossil bisons of European deposits to enable me to distinguish them specifically, but I am inclined to think that the appellation *B. priscus* has been made to include two and perhaps more species, and that a new designation is needed for the smaller American fossil bison. The selection of this name must be deferred till the arrival of more ample materials from Eschscholtz Bay (which I have reason to expect) reveals more fully the osteological peculiarities of the species, unless in the meantime the task is performed by some other naturalist.

Dr. Leidy's *Bison antiquus* rests as a species on a broken fossil horn figured in his Plate II. This fragment approaches in form and in dimensions to the horn-core of *Bison crassicornis* (Zool. of Herald's Voy. pl. xiii. f. 1 and 2), so as to beget a notion of their specific identity, though fuller evidence is needed to confirm it. Dr. Leidy's *B. latifrons* is a larger animal, but whether its great size be due to the age or sex of the individual whose remains have been disinterred, or be a characteristic of the species, requires to be proved. Dr. Leidy himself does not feel certain that *antiquus* is distinct from *latifrons.*

Of *Bootherium bombifrons* no crania have been discovered in the Eschscholtz Bay deposit, and the only bone that is at all likely to belong to *Bootherium cavifrons* is the *dentata* of *Ovibos maximus,* p. 25, Pl. XI. It is to be hoped that the discovery of a *dentata* of *cavifrons* in the United States will confirm or disprove this conjecture. It would appear that no authentic remains of *Ovibos moschatus* have been discovered in the valley of the Mississippi, and it is very likely that the animal never ranged southward of the Arctic Circle.

No remains of the restricted genus *Bos* have been detected in the tertiary deposits of North America, and there seems to be evidence sufficient to prove that even at the distant epoch of the occurrence of the drift phenomena, the fauna of that quarter of the world had assumed many of the characteristics which it still possesses, and which distinguish it from the fauna of Europe. In the European tertiaries we find numerous remains of at least three species of *Bos.*

Plate XVI.

Printed by P Reeve

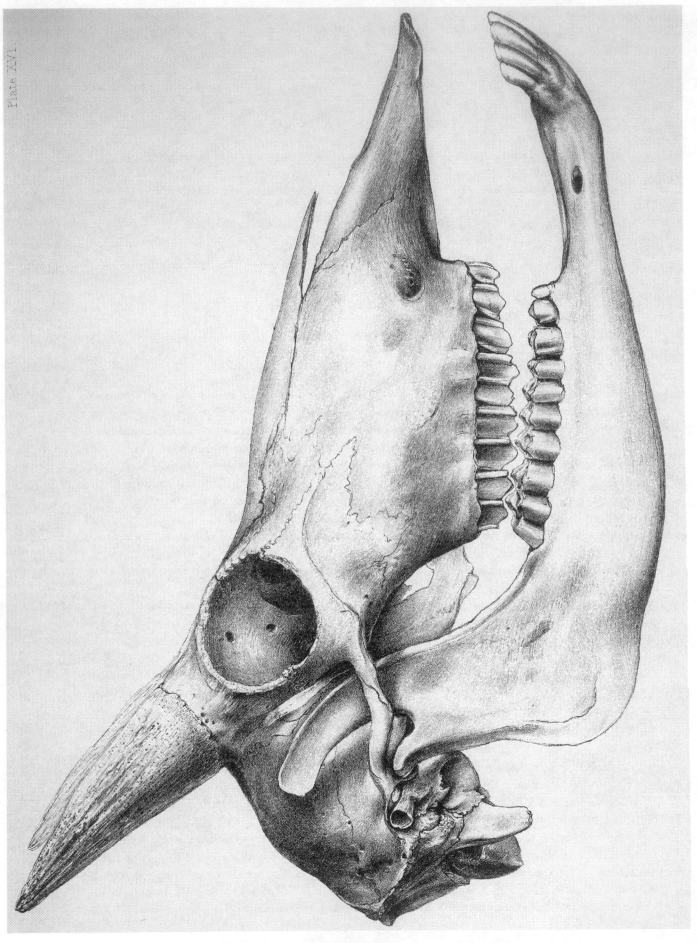

Drawn by W.Mitchell.

RUPICAPRA AMERICANA , *Skull nat. size.*

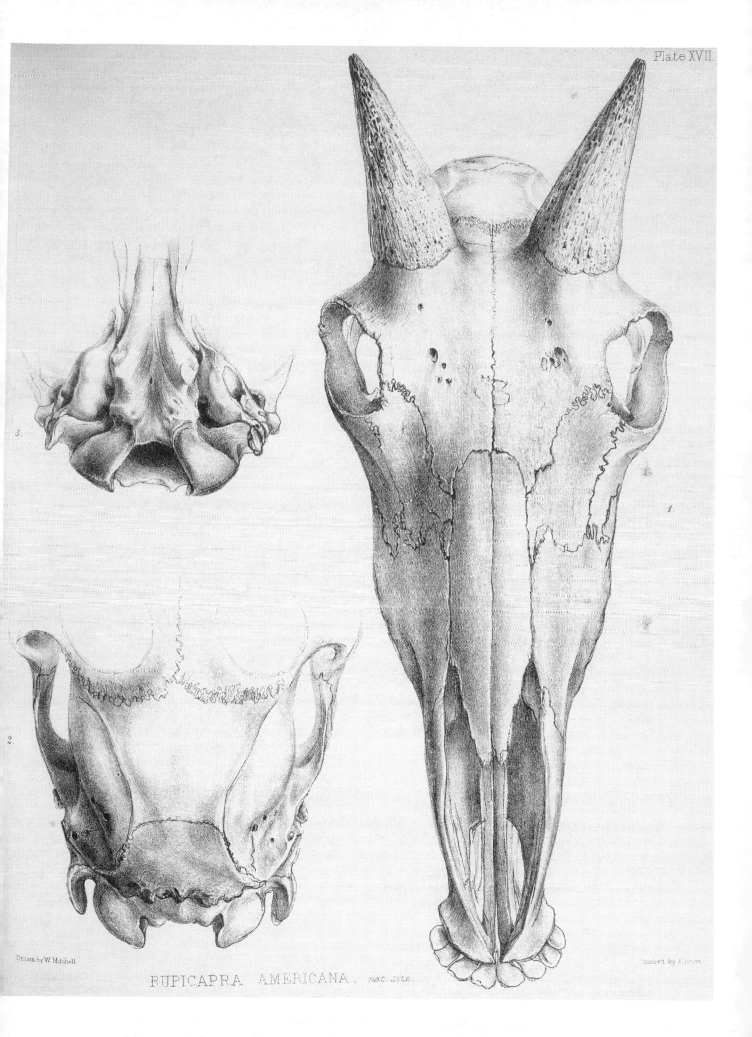

Drawn by W. Mitchell

Printed by

RUPICAPRA AMERICANA. *nat. size.*

Plate XVIII

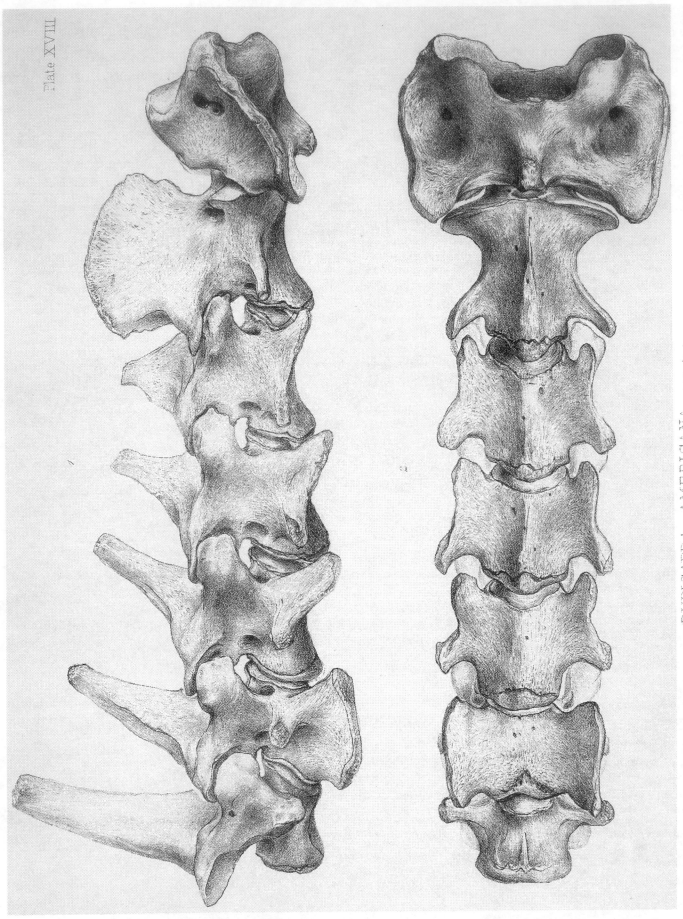

Printed by F.Reeve

RUPICAPRA AMERICANA nat.size)
Cervical vertebræ.

Drawn by W Mitchell

Plate XIX.

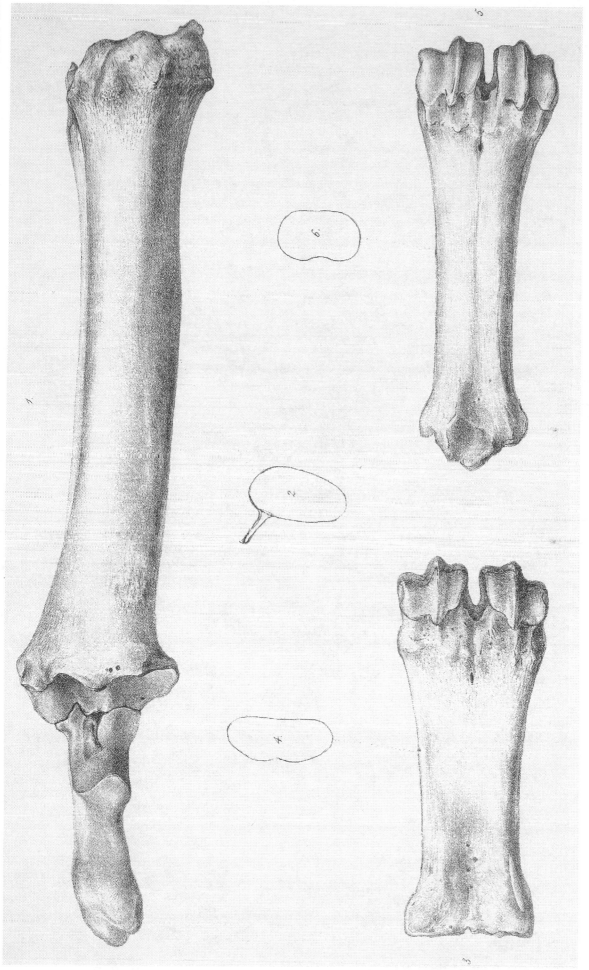

RUPICAPRA AMERICANA. (nat. size.)

1 Radius & ulna; 2. section; 3. Metacarpus; 4. its section; 5 Metatarsus; 6 its section.

Plate XX.

Printed by F.Reeve.

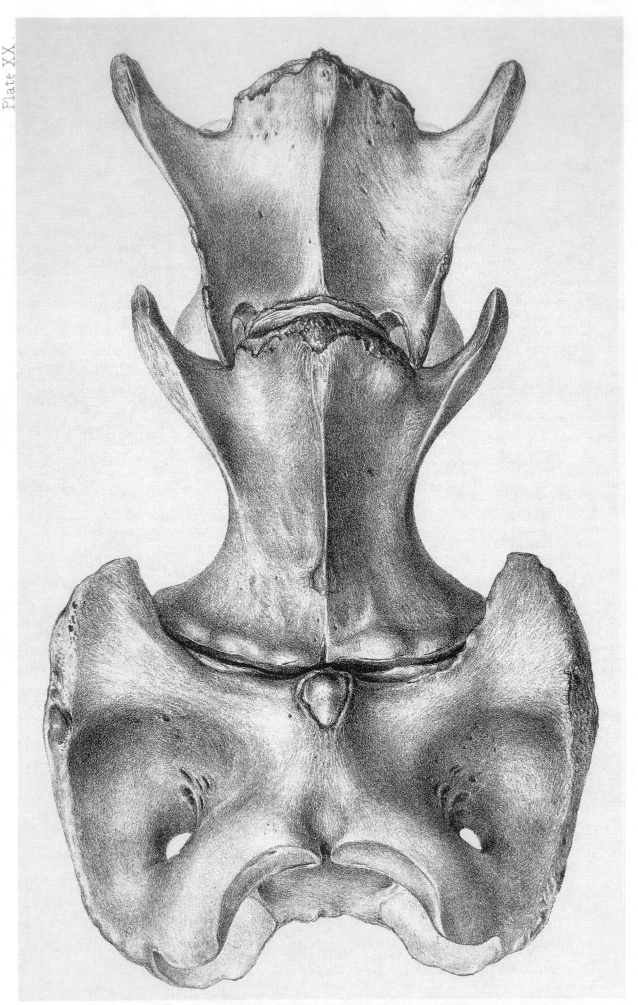

CERVUS ALCES, *(Muswa.)*

First three cervicals. natural size.

Drawn by W.Mitchell.

Plate XXI

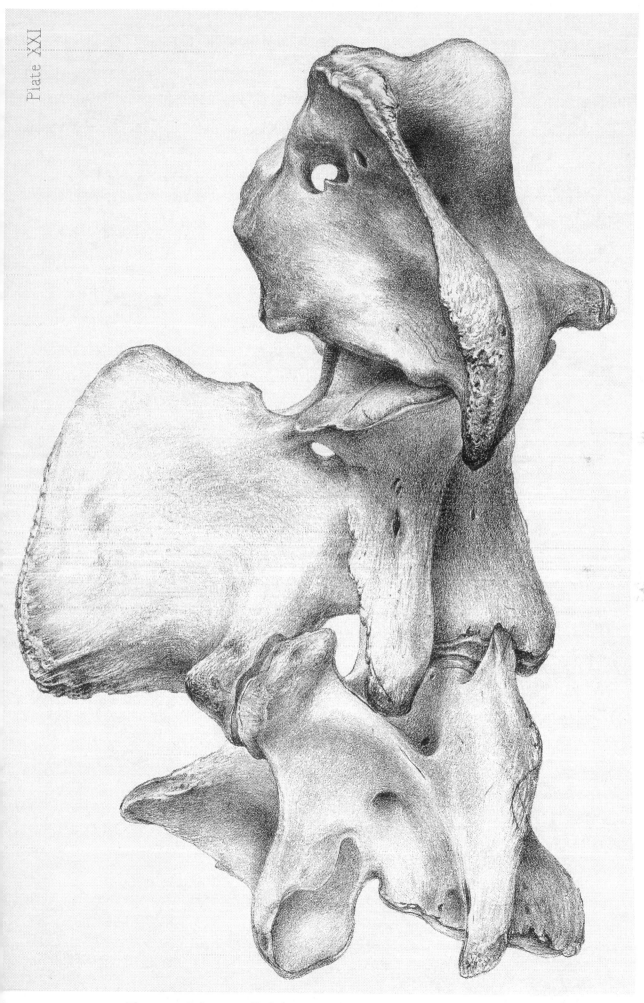

Printed by F.Reeve.

CERVUS ALCES (Muswa.)

First three cervicals. natural size.

Drawn by W Mitchell.

Plate XXII.

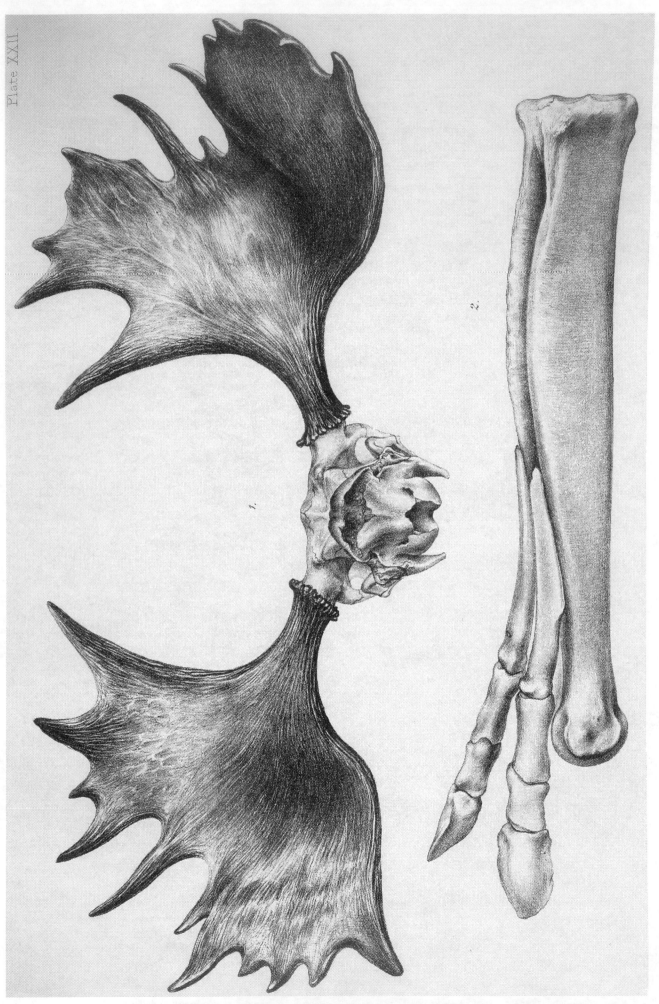

Drawn by W. Mitchell.

Printed by F. Reeve.

Fig 1 CERVUS ALCES *(Muswa)* ¾ natural size.
Fig 2 RANGIFER *(Tuktu)* Natural size.

Plate XXII.

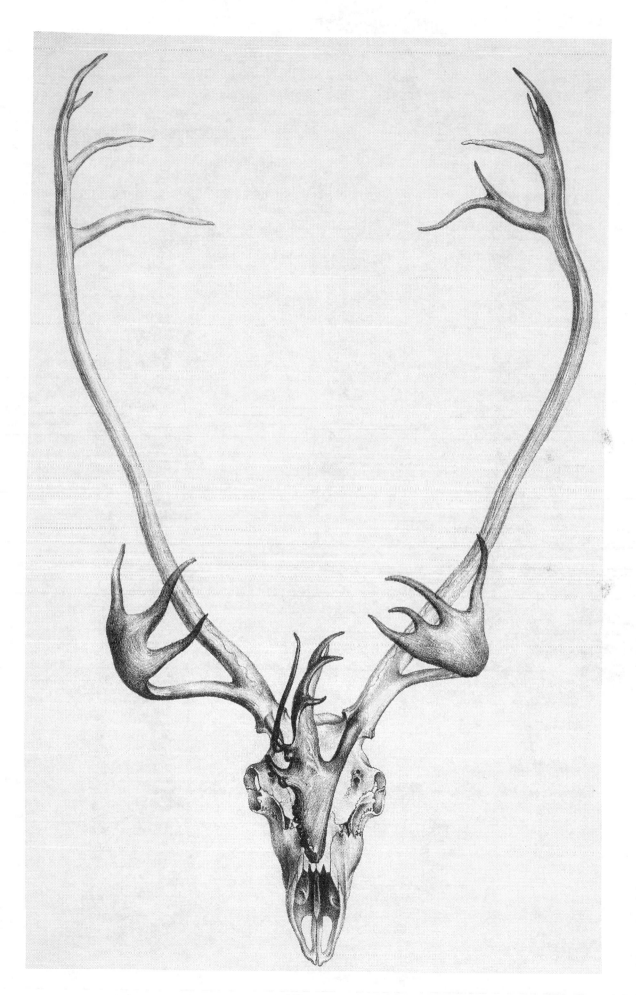

Printed by F. Reeve.

RANGIFER (*Tarandu*) ½ nat. size.

Drawn by W. Mitchell.

The material originally positioned here is too large for reproduction in this reissue. A PDF can be downloaded from the web address given on page iv

Plate XXIV

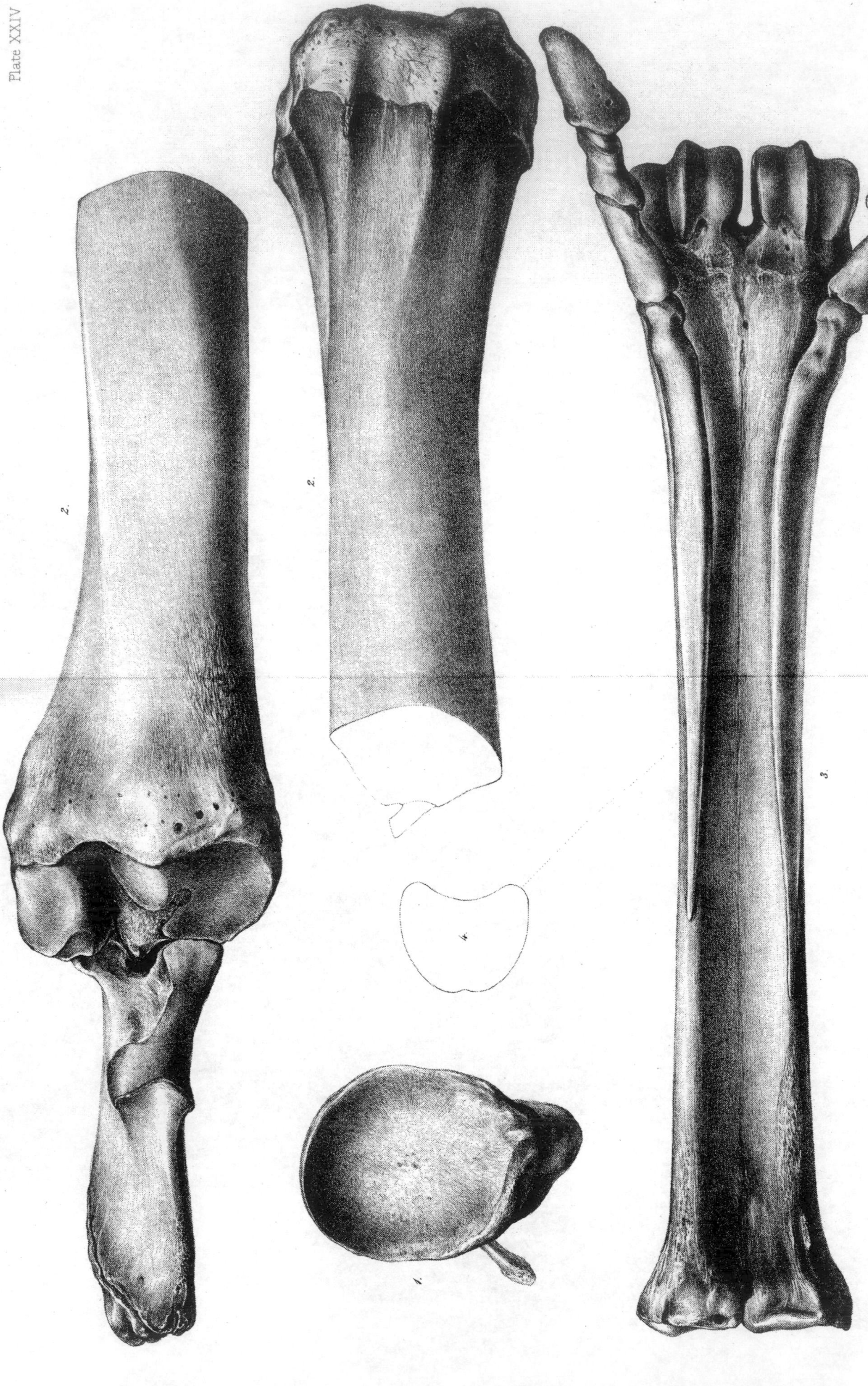

CERVUS ALCES (Muswa.)

Parts of the fore leg, natural size.

Printed by T.Reeve.

ZOOLOGY

OF THE

VOYAGE OF H.M.S. HERALD.

OSTEOLOGY—*Continued*.

MASTODON (?).

At page 102 it is stated that the scapula of the Mastodon does not exhibit the remarkable depression which characterizes the fragmentary shoulder-bones found at Swan River. Since I have (through the kindness of the author) had an opportunity of consulting Dr. Warren's excellent work on the *Mastodon giganteus**, I have discovered this assertion to be erroneous; a depression in the same part of the shoulder-blade of that species being noticed in the text by that gentleman, and figured in his large plate. The probability therefore is that the Swan River bones belonged to the *Mastodon giganteus*, and that the range of that species must be extended northwards in Rupert's Land to the fifty-second parallel of latitude, while the provisional geographical designation of *Elephas Rupertianus* must be expunged.

The depression in question was most likely designed to afford a firmer attachment to the central fasciculi of the *infra spinatus* muscle; and a similar one, though not so sharply defined, exists in the scapula of an Indian fossil elephant from the Sewalik hills, deposited by Dr. Falconer in the British Museum, as noticed at p. 102; but it is totally wanting in the several scapulæ of the Eschscholtz Bay elephants which are preserved in the British Museum and Haslar Hospital, the part in question being in them smooth and convex.

The error of my former notice above alluded to, arose from an inspection of Mr. Koch's skeleton of the Mastodon now in the British Museum, whose shoulder-bones exhibit no such depression. Neither is this character visible in two other scapulæ purchased by the same institution from Mr. Koch as bones of the Mastodon; all the four scapulæ having merely some roughness, but no hollow, in that part of the infra-spinal surface. From this fact one might be led to conclude that the concavity in question is merely an individual peculiarity, and does not occur generally in the species; but it is rare to meet a mere osteological variety so perfectly alike in form in the two limbs

* Description of the *Mastodon giganteus*, by John C. Warren, M.D. Boston, 1852.

U

as it is in our Swan River scapulæ, and, as we presume it to be, in both shoulder-blades of Dr. Warren's Newburgh Mastodon; for had it been otherwise, that accurate observer would have mentioned it. And the matter admits of another explanation. Mr. Koch's skeleton, when first brought from America for exhibition in this country, had its parts not only misplaced, but composed of the bones of more than one individual, there being at least five vertebræ too many in the spine. It may therefore be, that the two scapulæ now forming part of the skeleton of the British Museum Mastodon, and the two detached ones, are in reality bones of the American fossil Elephant, of which a cranium of great size was purchased by the Museum from Mr. Koch. Dr. Warren has shown that the *Mastodon giganteus* and the great fossil Elephant were coeval (*op. cit.* p. 142); and Mr. Koch may have dug up the remains of both animals from the same deposit. Not the least doubt rests on the authenticity of every part of Dr. Warren's skeleton of the Mastodon,—the account of its discovery and disinterment being quite clear.

The Swan River scapulæ belonged to an individual of intermediate size, between the Cambridge (Massachusetts) Mastodon and Dr. Warren's.

ELEPHAS PRIMIGENIUS.

Very recently a fossil skeleton of an Elephant has been discovered by Mr. Roderick Campbell, of the Hudson's Bay Company, on the sixty-first parallel of latitude, on the west side of the Rocky Mountain chain, near the sources of the Yukon or Kwichpack, and at an elevation by calculation of considerably more than 1500 feet above the ocean. The skeleton, when first found, was believed to be entire, but unfortunately, the Indians employed to disengage it and bring it home to the Fur Post, let it slip into a deep lake, where it now lies, with the exception of a tibia, which was recovered and brought to this country. This bone is of a bluish-white colour, has lost some of the rotular surface immediately beneath the knee-joint, and portions of the surface on its rotular and lateral aspects have scaled off; but is otherwise very perfect, the articulating surfaces of the knee and ankle joint being especially in good condition, and the popliteal side of the bone is entire.

On comparing the Yukon fossils with the smaller but more perfect *tibia* brought from Eschscholtz Bay by Captain Beechey, and now in the British Museum*, the popliteal aspects and distal articular surfaces are alike in both, as are also the fibular and medial aspects; though in this case the comparison is less satisfactory, from the surface having partly scaled off in the Yukon fossil. The circumferences of the proximal articulations are not perfect in either bone, but the parts which remain present no dissimilarities.

Dimensions of Tibiæ (by callipers).

	FROM YUKON.	ESCHSCH. BAY.
Length of the medial face of the shin-bone from the brim of the knee-joint to that of the ankle-joint	26·1 in.	18·8 in.
Length of the fibular face from the brim of the knee-joint to the proximal edge of the articular surface on which the distal extremity of the fibula moves	23·1	15·8

* Numbered 82·2 and 31A in the Catalogue of the British Museum.

	FROM YUKON.	ESCHSCH. BAY.
Length of the rotular face from the ridge between the articular cavities of the knee to the brim of the ankle-joint . .	26·6	18·8
Length on the popliteal face between the corresponding parts .	26·3	18·7
Circumference near the middle of the shaft where thinnest . .	11·9	10·4
Circumference of the distal end an inch and a half above the articular surface	19·5	

The fragment of a left tibia, No. 45, from Eschscholtz Bay, noticed in p. 16, belonged to even a larger animal then the Yukon fossil, since it measured 12·1 inches round at the middle. No. 54, a piece of the right tibia, measured 11·9 at the thinnest part of the shaft, corresponding with the Yukon one; but both the fragments from Eschscholtz Bay have their medial surfaces more convex than the two more perfect bones: this may be the effect of greater age.

REPTILES.

LOPHOSAURA GOODRIDGII (Gray).   *Goodridge's Basilisk.*

(*Lophosaura Goodridgii*, Gray, Ann. and Mag. of Nat. Hist. Dec. 1852, vol. lx. p. 438.)

(Plate XXVI.)

This Lizard, belonging to the section *Basilicinæ* of the *Iguanidæ*, was captured at Quibo by Dr. Goodridge, Surgeon of the Herald. Having no specimen of the Guiana Basilisk wherewith to compare it, I had referred it to that species on account of its general agreement with the descriptions given by authors; but Dr. Gray, of the British Museum, detected characters that separate it specifically from the previously described Basilisk, and are of value to raise it to a distinct genus or subgenus. On this point I defer entirely to his opinion concerning a group of animals that he has well studied, having indeed no means of judging for myself. The palate of *Basiliscus* is described by Cuvier as toothed, but in *Lophosaura* there is no vestige of a tooth on any part of the roof of the mouth.

Dr. Goodridge has recorded no particulars of the habits of the animal. The specimen was killed by a charge of small shot.

Description.

Form.—Its general figure may be comprehended more readily by the view of its profile, represented of the size of life in Plate XXVI., than by a mere verbal description. A peculiar character is given to it by its parietal crest, and high fin-like appendages on the back and tail. These, and the thickness of its body, do not indicate much activity unless in the water, but I do not know whether it actually takes the water voluntarily or no. The Guiana Basilisk is said to feed on berries.

The *head* is about one-fourth longer than its height at the occiput, exclusive of the crest; has flattish sides, and a somewhat trigonal coronal aspect. *Nose* blunt, whether it be viewed from above or laterally. In the profile, the upper eyelid forms a rounded projection, and the part lying between the eye and the crest a larger and more obtuse eminence. At the antinial border of this eminence, as viewed on the coronal aspect, there is a depression which is continued forward between the orbitar prominences to the flattish face. These eminences and depressions have very slight counterparts in the cranium. The *gape of the mouth* extends backwards past the eye to near the auditory opening, its commissure being slightly curved, concavely coronad. *Nostrils* formed externally by a small subtriangular or roundish opening on each side of the snout, near its end, and communicating with the roof of the mouth by a fissure on the lateral sides of each premaxillary. A slight swelling exists round the borders of the nasal orifices, and the rather large scales on the end of the snout are disposed in a peculiar manner and are porous. (*Vide* Fig. 2, Plate XXVI.) Each side of the *palate* is convex and smooth; and there is a median fissure, existing also in the skeleton, which runs iniad to the junction of the pterygoids. The internal orifice of each nostril opens between the smooth palate and the premaxillary. The *ears* are denoted by smooth, largely-exposed tympana, without prominent borders, situated behind the angle of the mouth and mostly above its plane. *Eyes* small, somewhat oblique, with tumid, movable upper and under eyelids, which, when closed, completely conceal the eyeball.

Crests.—Cranial crest blending in profile with the frontal eminences, but in strictness belonging to the parietals, and supported by a thin process of these bones, augmented by cartilage. Towards its edges and tip it becomes thin and flexible, and the latter overhangs the neck. The dorsal appendage is thinner, and the neural spines which support it, show through the scaly integument like the dorsal rays of a fish. This fin-like scaly membrane rises between the shoulders from the acute scaly ridge of the neck, and extends caudad to near the pelvis. Between it and the caudal appendix there are shorter neural spines, concealed in the recent animal by the muscles, but crowned by an acute scaly serrated crest, like the back of the neck, extending lengthwise for an inch and a half. Neural spines support the caudal appendix also; and a ridge which extends about two inches beyond it, on the tail, is also acute and serrated. Like the dorsal appendage the caudal one is somewhat scalloped on the border. The tail of this individual had been broken, and is reproduced nearly to the extent of six inches. A scaly wart shown in the figure marks the commencement of this reproduction. A short space between the broken part and the caudal crest is flat on its neural surface, which is protected by two rows of scales.

Feet.—The limbs are moderately long; the toes rather slender: those of the atlantal extremities being of moderate length, and without distinct webs at their bases. The third and fourth toe have a more distal origin than the others, or rather are further connected by integument at their bases, for in the skeleton the three middle toes arise side by side, and scarcely more distad than the lateral ones: the claw of the fourth toe however reaches the furthest, and in this respect the others come in the following order—the third, the second, the fifth, and the first; this last being not only the shortest, but also freeing itself from the carpal integument sooner than the fifth toe.

In the *sacral extremities* the toes are much longer, and at the same time of more unequal length; and they show a more marked difference in their origins in the recent foot, though in the skeleton the first four metatarsals are attached to the tarsus on the same line, the fifth only being more distant. In the recent foot the lateral or fibular toe separates from the sole nearest to the leg, and it is furnished on its inner side with a small fold of integument or basal web. The other toes have but a faint vestige of such a web, other than a scaly fringe to be afterwards mentioned. Next in origin is the mesial or tibial toe (great toe), and it is moreover the shortest and has the fewest joints. Then the second toe is disengaged, and equals the fifth in absolute length, but, having a more distal origin, passes it by nearly two phalanges. The third toe is considerably longer than the second; and the fourth is the longest of all, its length (without reckoning any portion of its metatarsal) being equal to that of the leg.

Teeth.—The teeth stand in a single crowded row on both jaws, in a groove, of which the inner border is deficient, or, as usually described, with their bases applied laterally to the thin edge of the jaw. Their stems are somewhat oblique, and are deeply channeled; their cutting edges convex, chisel-shaped, and tricuspid. The central cusp is much the largest and longest, the side ones being minute, and in the front teeth almost obsolete. All the cusps are acute. Each premaxillary supports four teeth, and each maxillary twenty-four; opposed to which, the mandible presents as many on each limb, or twenty-eight. The new teeth come up on the medial side of the old ones, causing the absorption of their bases.

Scales.—These vary in size and form on different parts, but almost all of them are either conical or keeled on the surface. They are smaller on the back and sides, than on the belly, tail, and outsides of the limbs, and smallest in the axils of the four extremities. A row of long, smooth, flat, quadrilateral scales, with oblique ends, borders each jaw, in place of lips. The succeeding scales on the maxillary are rather small, very irregular in outline, have uneven disks, and gradually assume an irregular obliquely conical form as they approach the top of the head. On the mandible the labial row is followed by one, and in some places by two, rows of large flat scales with irregular outlines. The pouch under the mouth is clothed by convex, rhomboidal scales, disposed regularly in oblique rows; while on the throat the scales are mostly pentagonal, and lie in transverse rows parallel to the fold that extends between the limbs of the mandible. They become much smaller, and roundish or oblong, and more convex as they near the pectoral region, on which they assume a different form and disposition.

On the eyelids the scales are minute, with two larger rows on the border of each lid. The transition from the larger rows to the smaller ones is graduated on the upper lid, but abrupt on the lower one. On the top of the head the scales are obtusely conical, and puckered or radiated, with an approach to the ridge-form, where they line the depression between the eyes and forward to the nostrils. The scales of the parietal crest undergo a gradual change from a conical shape to a somewhat keeled one, becoming larger and more oblong as they near the edge and apex of the crest. Behind the eye and on the neural and lateral aspects of the neck, they are smaller and proportionally more conical and uneven, showing also granular points on the connecting integument.

The scales of the back and sides are small, disposed in transverse rows, and for the most part are quadrilateral, with an oblique pyramidal disk, whose apex directed sacral is bluntish. On the belly the scales are also in transverse rows, but are more approximated: being moreover larger, flattish, smooth, and nearly square, with the sacral border slightly raised. On the thorax they are less regular in outline, and are disposed in two groups, which meet in a mesial furrow. The caudal scales are even larger than those on the belly, are ranged likewise in transverse rows, have a rectangular form, longer than wide, and are traversed longitudinally by an acute ridge. For some distance behind the vent the scales are flattish and smooth, but as they recede further the median keel gradually assumes more prominence, and, from the correspondence lengthwise of the keels of the successive rows, five longitudinal raised lines are produced on the convex under surface of the tail. The keeling of the scales is fainter, and the longitudinal lines consequently less conspicuous, on the lateral parts of the tail. The dorsal and caudal appendages are clothed with larger scales than any other parts.

Most of the scales covering the *limbs* are rhomboidal, and set angularly, with the median ridge parallel to the axis of the limb. On the palms the scales are small and obtusely conical, imparting roughness to the surface. And the toes also of the *atlantal extremities* have a degree of roughness produced by the prominence of the edges of the lateral scales: their anconal surface is covered by very obtuse scales, each traversed by three or four obtuse ridges, while the thenal surface is furnished with scales which are smoothly rounded, forming each a transverse prominence, that admits of the toe grasping any object closely. The *claws* are short, compressed, curved, and acute.

An obtuse conical point on each scale gives roughness to the soles, as in the palms; but the plantar surface of the toes is more complicated than in the atlantal extremities, each toe having several rows of rounded scales, more or less obliquely and irregularly disposed. The lateral scales on the ulnar sides of the first, second, third and fourth toes project, forming an acute border; the fifth toe has no distinctly formed border, but it is serrated on both sides by the prominence of its scales. The *claws* are a little longer and more acute than those of the atlantal extremities. There are *no pores* on the thighs.

Colour.—After long maceration in spirits the specimen has a dark grey colour on the dorsal aspect, and is nearly white on the belly and inner or ventral aspect of the limbs. The sides of the body and base of the dorsal appendage are traversed by six or eight oblique, irregular, deep black bands; the thighs are spotted with black, and many of the dorsal scales have black disks. Except where the oblique bands encroach a little on the dorsal appendage, neither that nor the caudal one have (or at least retain) any black markings. The cuticle having partially peeled off on the sides and belly, there has probably been some obliteration of colour.

Dimensions of a mature specimen after two years' maceration in spirits.

Length from the tip of the nose to the anal orifice	7·9 inches.
Length from the anal orifice to the tip of the tail	16·0
Total length	23·9
Length of head to the apex of the crest	3·3
Length of atlantal extremity from the axilla to tip of the longest toe	3·3
Length of the thigh of sacral extremity	3·3
Length of the leg	2·0
Length from extremity of the heel to the origin of the fourth or largest toe	1·3
Length of the longest toe, including its claw	2·0
Length of attachment or base of dorsal appendage	3·8
Height of tallest ray of the dorsal appendage	1·8
Distance between the dorsal and caudal appendages	1·3
Length of attachment of caudal appendage	7·2
It greatest height	1·5

Osteology.

The skull is a cone of four sides, the nose and tip of the mandible forming the blunt apex, while the irregular base is formed by the occiput, tympanic bones, and prolonged angles of the mandible; frontal having nearly the form of the letter T, the cross piece being applied to the parietals by a perfectly straight transverse suture immediately behind the orbits. Articulated to the lateral point of the cross piece of the frontal, and to the contiguous process of the parietal, is a short pedicle of the post-frontal. This bone constitutes the fifth part of the circumference of the orbit, and expands atlantad into a thin broadly lanceolate plate, which is applied splintwise to the junction of the squamosal and malar, lies in the same plane with them, and forms a greater part of the breadth of the zygomatic arch than these two bones do conjointly[*]. The squamosal is narrow in comparison, and does not quite reach the point of the rather narrow malar, the arch being perfected by the interposition of a small part of the post-frontal. In *Oreocephalus* the squamosal is shorter and broader, and the malar joins it more atlantad.

The frontal is moderately concave on its coronal aspect, there being only a slight elevation of the edges of the orbits, and a still smaller prominence of the transverse fronto-parietal suture. Throughout the

---

[*] In the crocodiles, the descending branch of the post-frontal is a roundish stem lying mesiad of the disk of the malar, and connected with its interior surface, quite clear of the squamosal.

whole length of the pariotals they rise gradually towards the site of the sagittal suture, to form a thin elevated mesial crest, which is prolonged sacrad, so as to overhang the second cervical, and is further extended at its tip by a membranous border. At their inial angles the parietals flange out laterally with a partial twist, and overlie the mastoids. The angles of the skull at its widest part (excluding the mandible) are formed by the articulations of the squamosal with the mastoid and parietal—the squamosal covering the ends of both laterally. These articulations project iniad of the occipitals.

Mandible.—The part of this bone which is armed with teeth is equal in length to the remainder. Just behind the teeth a strong coracoid process rises coronad, and plays on the inner surface of the malar, while a process springing from the articular part, and having an obtuse point, projects to a considerable distance iniad of the joint, with an inclination also coronad. Opposite to this process, and directed basilad, there is a small acutely angular projection, whose homologue in *Chlamydophorus* is large and strong, is but slightly developed in *Craneosaura*, and is not perceptible in *Oreocephalus*.

Vertebræ.—In preparing the skeleton, some of the sternal ribs have given way, so that the exact number attached, directly or through the medium of cartilage to the sternum, cannot be made out. There are twenty-four vertebræ anterior to the sacrum, two sacrals, and thirty caudals, exclusive of the reproduced part of the tail, which occupies the length of about twelve more caudals, making fifty-eight vertebræ in all.

The first four cervicals have no pleurapophysis, but have projecting median hypapophysial ridges. A short rib is distinctly articulated to the diapophysis of the fifth cervical, and the sixth and seventh have longer ribs, whose ends pass within the clavicle, but do not appear to have any attachment to the sternum. The hypapophysial ridge is obsolete in the seventh cervical.

Four ribs are articulated directly to the sternum, and three others by the intervention of a common branching cartilaginous rib. The succeeding ribs gradually shorten, but continue to be articulated to the vertebræ down to the sacrals, there being in all twenty pairs of ribs, cervical, dorsal, and lumbar.

None of the vertebræ beyond the fifth or sixth cervical have a hypapophysial ridge, until we reach the third caudal, which, with those that follow it down to the sixteenth or eighteenth caudal, has an arched hæmapophysis and spine as in fishes, becoming smaller as it is more distal, and being minute on the sixteenth, seventeenth, and eighteenth.

The *neural spines* from the second to the sixth cervicals are low and equal; that of the seventh is a little longer, and they lengthen gradually to the ninth and tenth dorsals, which are the longest; decreasing again to the fifteenth. Five of the more proximal dorsal spines are strong, and are augmented by a thin border directed atlantad; but more distal ones become successively more and more filiform, and all are strengthened at their bases by a thin plate standing atlantad. The last two dorsals and the two sacrals have short neural spines; but the succeeding processes lengthen gradually, until in the third or fourth caudal they rise above the level of the back, and become the ribs of the caudal appendage. The spines supporting the caudal appendage are mostly nearly linear, tapering little, and are all compressed. The appendage may be said to end with the twenty-second caudal, whose spine scarcely rises beyond the general surface of the tail. After the twenty-seventh caudal there are no neural spines. The reproduced part of the tail has an osseo-cartilaginous centre, not divided into distinct vertebræ.

An episternal, shaped like the letter T, has its cross-bar embraced by the ends of the clavicles, which meet over its atlantal surface. Its descending stem lies on the ventral surface of a large rhomboidal sternum, or ossified union of the sternal ribs. The two proximal sides of the rhomb are wedged in anglewise between the large coracoids; and the styliform lateral end of the clavicle lies along the edge of the scapula, to be joined by its point to the supra-scapula. This last-named bone is large, and is ossified in a white granular way, like the sternal piece. Above the ulnar olecranon there is a small sesamoid, such as Professor Owen has noticed on the Bat ('On the Nature of Limbs,' p. 24).

Of the *metacarpals* the third is the longest, the fourth nearly equals it, and the fifth is the shortest.

The first toe has two phalanges, the second three, the third four, the fourth and longest five, and the fifth three.

Sacral Extremities.—Pelvis strong and complete. The ilium, a strong, linear, compressed bone, is attached to the ends of the diapophyses of the two sacrals, projects a little dorsad of these processes, and also emits an obtuse projection at its distal end, immediately atlantad of the acetabulum. The pubals stand obliquely sternad and atlantad of the last-mentioned projection of the ischium, expand into a triangular shelving plate, and meet in a narrow symphysis. The ischia also complete the pelvic arch by symphyses, from which two styliform median processes issue, one atlantad, dividing the large under-opening of the pelvis into two foramina ovalia, the other pointing caudad.

Instead of a single *patella*, there exists a small sesamoid at the head of the fibula, and a still smaller one on the opposite side of the tibia. Of the *metatarsals* the fifth or fibular one is the shortest. The others become longer in succession from the first to the fourth, which exceeds the third very slightly. The first toe has two phalanges, the second three, the third four, the fourth five, and the fifth three.

Dimensions of the Skeleton.

Length from the tip of the premaxillaries to the inial end of the bony parietal spine or crest	2·5 inches.
Length from ditto to the middle of the occipital ridge	1·8
Length of the bony orbit	0·69
Height of ditto	0·51
Length of the mandible, including the projection of the articular piece	2·22
Breadth of the skull at the malar plates	1·1
Length of the humerus	3·31
Length of the radius	0·9
Length of the fourth metacarpal	0·34
Length of the four phalanges and claw of that toe	0·88
Length of the femur	2·15
Length of the tibia	1·9
Length of the fourth metatarsal	1·05
Length of the four phalanges and claw of that toe	1·85

CRANEOSAURA SEEMANNI (Gray).

(*Ptenosaura Seemanni*, Gray, Ann. and Mag. of Nat. Hist. for Dec. 1852, No. 60, p. 438.)

Ptenosaura or *Ctenosaura* was preoccupied, and was changed by Mr. Gray, as above. *Th. κρανος, galea.*

(Plate XXV.)

Description.

Body oblong-oval, tapering gradually into the slender tail, rather compressed, being nearly one-third higher than wide, and having the summit of the back acutely ridged by the prominence of a series of strong compressed triangular scales.

The *head* is not very dissimilar to that of *Lophosaura*, the difference being principally in the form of the parietal crest. Snout obtuse in profile, and also as seen on its coronal aspect. Eyebrow prominent, covered with somewhat pentagonal scales, small, irregular, and mostly obscurely ridged. One or two rows of larger scales encircle the eyebrow, the lateral part of the circle being formed of compressed ones that constitute an acute fold projecting over the upper eyelid, and the rest lining the median furrow of the face. The flattish region of the snout bounds this furrow anteriorly, and the front of the parietal crest rises at its inial end. At its base the crest consists of a roundish mass of muscle, supported by a thin central plate of bone, and projecting above the back of the neck; from this there rises abruptly a very thin elastic crest having an even border, and, when the base is included, an oval outline. The whole of the crest, with its bulging base, is covered by pentagonal scales, which are smaller and more or less obscurely ridged on the fleshy part, but having throughout smooth disks and no elevated edges. Each nostril is a round orifice in a membrane, has a lateral aspect, and is situated much nearer to the end of the snout than to the eye. Both upper and under eyelids are movable, and cover the eyeball completely; their scales are small and rounded, with a row of larger compressed marginal ones. *Tympanum* oval, smooth, and rather large, without border or perforation.

Commissure of the *mouth*, having a wavy curve, extending to near the tympanum. Its borders, above and below, are formed by oblong, transversely convex scales, which on the mandible are flanked by a wider series, there being however towards the corner of the mouth a triangular patch of smaller scales interposed between the rows of the larger ones. Smooth, very convex scales, having their distal edges most prominent, cover the gular pouch and the entire space between the limbs of the mandible. On the swelling fleshy angles of the mandible these scales gradually acquire a flatter surface and more rhomboidal outline, and some minute ones are interposed among them. The gular pouch is prominent, though not capacious, and several folds of scaly integument descend towards it, from before the shoulder.

Teeth.—An even row of tricuspid teeth edges the upper and under jaws, the middle cusp being nearly semicircular, thin-edged and more projecting, and the lateral ones minute; the teeth generally being very similar to, but rather wider, than those of *Lophosaura*. There are 20 teeth on each maxillary, 5 on each premaxillary, and 26 on each limb of the mandible. The premaxillary teeth, and the mandibular ones opposed to them, are narrower and more pointed than the others, most of their lateral cusps being obsolete. There are also a series of 3 or 4 minute acute teeth on each palate-bone along the border of the mesian palatine fissure. No such teeth exist in our specimen of *Lophosaura*.

Scales.—Small, roundish, irregular scales cover the back of the neck, while larger and more rhomboidal ones protect the throat. On the back and upper parts of the sides the scales are small, and have for the most part square disks, though some are rhomboidal and others triangular, with more or less of irregularity. On the sides many are perfectly rhomboidal, others roundish in outline and convex. They are rhombs on the breast in part, in part irregular, to suit the convergence of the rows towards a mesial furrow on the sternum. Somewhat larger rhomboidal scales, disposed in irregular transverse rows, protect the belly, and the same transverse and uniform disposition prevails on the tail, but most of the caudal scales have their disks partly crossed lengthwise by a smooth, elevated line. Towards the end of the tail the ridges of the scales become more and more distinct, until at length they form four very acute and continuous longitudinal lines or keels. On the outsides of the shoulders and fore-arms the scales are rhomboidal or obtusely semi-lanceolate, with a median ridge, and are arranged so that the ridges produce prominent lines of various lengths. This arrangement is also very conspicuous on the front of the thigh and leg. Small convex scales protect the axillæ of the thighs and shoulders.

Extremities.—There is no distinct fringing of the *fore-toes*, though the arrangement of the scales produces an approach to it; the scales on the thenal surface of the toes being transverse rolls, while the lateral ones are more or less compressed and keeled, and project slightly. The thumb and fifth toe separate from the palm opposite to each other, then the second toe becomes detached, and the third and fourth are connected a little further. Nothing like a web exists between the atlantal toes.

The first four toes of the *sacral extremities* have distinct borders on their fibular sides formed of thin scaly integument, which is serrated by the prominence of the points of the scales. It is conspicuous on all these four toes, but is widest on the fourth. The fifth toe is serrated on both sides by compressed acute scales, which are narrower and stiffer than those bordering the other toes. The fifth toe separates first or highest on the sole; then, at a considerable distance after it, come in succession the first, second, third, and fourth, standing as it were *en échelon*. The fourth is the longest toe. On both feet the claws are compressed, curved, and acute.

Colour.—The original colours of the lizard have changed by maceration for upwards of two years in spirits. At present the most general tint is an olive, varied in a small tessellated pattern by numerous pale green and white scales. Broad rings of olive and pale green alternate on the tail, and the pale tint is disposed in chevrons on the outer sides of the extremities.

Dimensions

of a mature specimen after maceration in spirits (by callipers).

Length from the tip of the nose to the anal orifice	5·3 inches.
Length of fragment of the tail represented in the plate	15·5
Length of head to extreme inial margin of the crest	3·26
Height of crest above the upper border of the tympanum	1·9
Length of the crest	2·16
Length of the commissure of the mouth (laterally)	1·74
Length of atlantal extremity from the axilla of the shoulder to the top of the longest toe	3·3
Length of the thigh	2·46
Length of leg	2·42
Length of the tarsus and longest toe and claw	3·4
Height of the body midway between the atlantal and sacral extremities	2·0
Greatest width there	1·6
Circumference there	5·3

Osteology

of a smaller specimen than the one described above.

In its general form the skull has a strong resemblance to that of *Lophosaura*, the nasals and frontals being however narrower, and the thin mesial edges of the orbit more elevated. The transverse fronto-parietal suture forms a more prominent ridge, that reaches from the inial canthus of one orbit to that of the other; a minute notch in the middle of the ridge receives the commencement of the parietal spine or crest as into a mortice, and the inial extremity of the crest is prolonged so as to overhang the third cervical. The form of the bony crest is oblong-linear, the oval of the crest in the recent animal being completed by thin cartilage.

The *malar plate* of the *post-frontal* is not so wide as in *Lophosaura*, but the zygoma is built up in the same way. The *mandible* is similar to that of its homologue in the species just named; the coronoid process is alike in the two, as is also the prolongation of the articular piece beyond the joint.

The *pelvis* is destitute of the sub-ossified styliform process, which in *Lophosaura* projects caudad from the symphysis of the ischia. There are in all 23 vertebræ in the spinal column, anterior to the two which represent the sacrum and support the ilia. Of these, the first 3 cervicals have no articulated ribs, all the rest have, and 7 are connected to the sternum by long and slender cartilages. The lumbars have short articulated ribs. The dorsal spines are a little taller towards the middle of the series, but not abruptly, the whole series being nearly even. On the tail they gradually lower to a mere point, but in this specimen nearly half the tail has been reproduced, and is semicartilaginous, without separation into distinct vertebræ. The inferior caudal spines are like those of *Lophosaura*.

The olecranon is a distinct bone from the ulna, as in *Lophosaura*, and there are two minute sesamoids in the lateral parts of the knee-joint, but no other patella.

Dimensions of the Skeleton.

Length of skull from the tip of premaxillary to the end of the parietal crest	2·2 inches.
Length of ditto from ditto to middle of occipital ridge	1·6
Length of the bony orbit	0·62
Height of ditto	0·52
Breadth of the skull at the malars	0·9
Length of the mandible to the tip of the process of the articular piece	1·9
Length of the humerus	1·8
Length of the radius	0·86
Length of the fourth metacarpal	0·33
Length of the four phalanges and claw of that toe	0·75
Length of the femur	2·05
Length of the tibia	1·82
Length of the fourth metatarsal	1·05
Length of the four phalanges and claw of that toe	1·82

GECKO REEVESII (Gray, Griffiths, Cuv. ix. p. 48).

Cat. Br. Mus. Lizards, p. 161.

(Plate XXVII.)

Form.—Body elongated; head large; tail tapering, roundish and pointed, and not far from equal in length to that of the head and body combined. Limbs short and thick. Toes dilated, lamellar beneath.

Head, when viewed on the coronal aspect, egg-shaped in outline, as broad posterior to the eyes as the broadest part of the body, becoming gradually narrower towards the snout, which is obtuse and rounding off more bluffly into the neck: in profile, flatly convex. The somewhat prominent eyelids produce a mesial furrow; the coronal surface of the cranium is flat transversely or even somewhat concave, and there is also a slight concavity before the eyes, but the end of the nose is convex. The height of the head before the auditory openings does not exceed half its width, and the temples round off into the flat vertex.

The *nostrils* are small round holes, looking antiniad and laterally, one on each side near the tip of the snout. On the *face* and *top* of the *head* the scales are small and convex, and differ little from each other in size, some however being polygonal with rounded corners. The same kind of scales protect the upper eyelid down to the margin, which is formed of a row of large rectangular flattish scales, backed by two rows of minute, irregular ones. The *upper eyelid* is movable, and covers the upper part of the eyeball, but tapers away at each canthus, and the under eyelid is obsolete. *Tympanum* deeply sunk in an oblique, oval cavity, situated behind the widest part of the head. The edges of this cavity are full and rounded, and are protected by scales that are smaller than those of the top of the head.

Commissure of the *mouth* extending to the posterior part of the eye, with a sudden, short, upward curve at its angle, from whence to the ear there is a fold of integument. A row of distant, somewhat com-

x 2

pressed, subulate, but not very acute teeth, arms the premaxillaries, maxillaries, and mandible.   There are none in the interior of the mouth, but the gular half of the tongue is rough with fringe-like papillæ. Both jaws are edged by a row of large, nearly uniform, square or rectangular scales, and at the symphysis of the mandible the marginal row is succeeded by somewhat smaller polygonal or roundish ones, each successive row diminishing rapidly in size, until lost in the roundish or oval small scales which cover the rest of the skin between the limbs of the mandible.   A fold of integument runs transversely from one ear to the other, but there is no distinct gular pouch.

On the *breast, belly*, and *inside* of the *thighs* the scales are again somewhat larger, have a rhomboidal or roundish outline, and a flat disk.   On the *back* and upper surface and sides of the tail they are roundish, like those of the crown of the head, varied by five or six longitudinal rows of smooth, convex, larger and prominent ones, disposed rather remotely at pretty equal distances, and giving a tubercular appearance to the back.   The tubercles, in a less conspicuous manner, are continued down the tail.   On the under surface of the tail a series of circular bands are visible, each composed of two pairs of larger rectangular scales, with a small rhomboidal scale occupying the angle where the two pairs meet.   The rest of each band is filled up on the side and tip of the tail with smaller scales in five or six rows, and with the rows of prominent ones as above mentioned.   Anterior to the anus there is a transverse chevron of from 18 to 20 porous scales, meeting at an angle in the mesial line.   Our two specimens, being females, have no pores on the thighs.

The *toes* spring from the palms and soles in a radiated manner, the inner one being the shortest and also destitute of a claw, and the penultimate one the longest.   All are edged on both sides by a row of somewhat compressed, blunt, tiled scales (fig. 7, 8), and their under surfaces are furnished with a series of soft flexible laminæ, represented in figures 5, 7, and 8.

The *colours* of the specimens have disappeared through maceration in spirits, but the markings represented in the Plate are still discernible.

Dimensions.

Length from the symphysis of the mandible to the verge of the anus	4·4 inches.
Length from the verge of the anus to the tip of the tail	4·5
Length of head	1·74
Length of atlantal extremity	1·82
Length of sacral extremity	2·1
Greatest width of the body	1·2

The specimens were procured at Canton by Dr. Seemann.

Osteology of a Gecko.

We have not seen a skeleton of the *Gecko Reevesii*, but subjoin some particulars of the osteology of a very closely allied species which is common at Tenasserim, on the coast of Siam, and which is most probably the *Gecko verus* of Dr. Gray's Catalogue of the species in the British Museum.

The skull, and indeed the entire skeleton, differ much in character from those of the *Iguanidæ*.   The cranium is broad and flat on its coronal aspect, and has a short, wide, and obtuse nose.   A single premaxillary, which ends iniad in a lanceolate tip, that enters a notch between the nasals, supports 11 somewhat compressed and slightly tapering teeth, having rounded chisel-shaped crowns.   A row of eight foramina pierce the maxillary above its dentary border, and a few small apertures are scattered over its disk.   There are from 30 to 31 teeth in each maxillary, a little larger than the premaxillary ones, but otherwise similar.

Each limb of the mandible is armed with from 32 to 35 teeth like the upper ones. All the teeth lean against and are adherent to an exterior parapet of their respective jaws, their roots being exposed in the skeleton, and many of them more or less eroded by successional teeth in various stages of growth. The mandible has a low inner parapet, in form of a smooth, obtuse ridge, on the mesial side of the teeth and a little distant from them, constituting with the outer one a wide groove, in which the teeth stand. Such a groove exists only in the inial half of the maxillary. A series of six foramina pierce each limb of the mandible.

Only a small part of the border of the orbit is formed by the frontal. The interorbital portion of this bone is equal in breadth to the pair of nasals, and is single, the sagittal suture being obliterated; while the transverse processes, which in the preceding *Iguanidæ* form the transverse head of the letter ⊤, are short and acute, their points exactly coinciding with the acute antinial corners of the broader parietals. These corners of both the bones are embraced equally by the deeply crescentic post-frontal, which emits only a short corner laterally, and leaves the outer fourth part of the circumference of the orbit to be completed by ligament. Another fourth part is formed by the pre-frontal, which lines part of the maxillary and more of the frontal, approaching on the edge of the latter bone the tip of the post-frontal to within a third of its own length, this intervening part being the only portion of the margin contributed by the frontal proper. The lachrymal is confluent with the maxillary, so as not to be distinguishable from it on the edge of the orbit. A slender styliform malar lies splintwise on the inial end of the dentary process of the maxillary, is partly confluent with it, and forms one-fifth of the border of the orbits. The extreme end of this part of the maxillary descends a little basilad beyond the last tooth, is connected mesiad with a corner of the pterygoid, and affords a lateral facet, against which the well-developed coronoid of the mandibular plays: but the zygomatic arch, so conspicuous in the *Iguanidæ*, is deficient in *Gecko*.

The inial point of the cranium is formed on each side by a long slender process of the parietal joined splintwise to the mastoid, and supported by a transverse process of the exoccipital, between which and the parietal process there is a pretty large opening*. A broad tympanic, concave iniad, and convex on the other surface, descends from the mastoid, and runs obliquely between basilad and antiniad, to furnish at its ex-

* The confluence of some bones and the obscurity of the connections of others, hidden under the ligaments of a natural skeleton, may have caused me to commit an error in the designation of the mastoid in the *Gecko*, so that a few words of explanation may not be unnecessary.

In *Craneosaura* a short, stout mastoid is adherent to the inial surface of the elongated, twisted corner of the parietal, and, in conjunction with the expanded inial end of the squamosal, furnishes a joint for the head of the tympanic, to which lateral support is given by the transverse process of the exoccipital. In the *Lophosaura* the mastoid is considerably smaller, and is more confluent with the twisted angle of the parietal, so as scarcely to appear distinct. The passage between it and the transverse process of the exoccipital is rather small.

In *Oreocephalus* the squamosal has a rhomboidal lateral surface, is joined to the malar not splintwise, but by a less oblique suture, and its broader inial extremity articulates distinctly above with the angle of the parietal, and constitutes most of the joint for the head of the tympanic,—the lateral part of the mastoid being small in comparison. The membranous walls of the auditory, opening with some ossified points, lie, in the skeleton of this lizard, basilad of the squamosal and mastoid, and between them and the elongated process of the articular segment of the mandible.

In *Gecko* I consider the squamosal to be wanting, and the mastoid to usurp its function of forming the fore part of the tympanic joint. A slender process of the mastoid lies splintwise on the antinial side of the very slender parietal process; while in the three lizards above noticed, the mastoid adheres to the inial side of that process. The principal part of the tympanic joint is formed in *Gecko* by the process of the exoccipital, which stands out obliquely laterad and iniad. The pterygoids are built up into a very beautiful internal framework connected with the presphenoid, giving lateral support both to the tympano-mandibular arch and also to the maxillaries and palatines.

tremity a condyle for the articular part of the mandible. A concave truncated process of this bone projects iniad of the joint, but not so far as the mastoid, though it still forms a strong lever to the jaw. The gape of the mouth is large. There are no teeth on the palatines or pterygoids; and the sclerotal coat of each eye is supported by 15 bones.

There are 26 vertebræ anterior to the pelvis. Of these the first two seem to have no articulated pleurapophysis. A short rib is attached to the 3rd, and rather stout ones to the 4th and 5th. The 6th and onwards to the 13th or 14th complete the arch, by the intervention of the sternum and its ribs or cartilages. Six succeeding vertebræ have long but floating ribs, after which the pleurapophyses rapidly shorten. The 27th and 28th vertebræ support a short, stout process lying transversely, whose broad end is articulated to the oblique ilium near its middle. Shelving pubals complete, by their symphysis, the pelvic arch anteriorly, and the union of the ischia completes it posteriorly, there being a wide round opening between the two symphyses. The tail of the specimen has been mutilated, but no hæmal canal seems to have existed in the caudals anterior to the fourth from the sacrals.

Several sesamoids exist in the knee and elbow joints, and there is a conspicuous one above the olecranon. The femur is considerably longer than the tibia, and the humerus also exceeds the radius in length.

ANNIELLA PULCHRA (Gray).

Anniella pulchra, Cat. Br. Mus. Reptiles, p. 88; Annals and Mag. Nat. Hist. Dec. 1852, vol. x. p. 440.

(Plate XXVIII.)

This reptile, of the family of the *Scincidæ*, was discovered in California by Dr. Goodridge, the able and industrious surgeon of the Herald. It belongs to the tribe of *Siaphosinæ* in Dr. Gray's arrangement, and, being the first of that group which has been detected out of Australia, its discovery is a matter of interest to the student of the geographical distribution of forms. It differs from *Soridia* in wanting the rudimentary sacral limbs, but otherwise has a very close resemblance indeed to that genus.

Description.

Form.—Like *Soridia*, this Skink is worm-shaped, and in both, the tail is capped by a single convex scale, which is more obtuse in *Anniella*. The head is alike in both, and is smaller than the fore part of the body. Eyes very small, with a single row of scales intervening between the posterior edge of the mouth and the under eyelid, which is itself a single narrow scale (fig. 2). Three large, convex, nearly rectangular scales protect the maxillary, anterior to which a rhomboidal one, pierced by the minute nostril, also forms part of the convex margin of the mouth; and the end of the snout is covered by a single convex pentagonal scale, which fits in between the pair of nasal ones (fig. 3). This initial scale is followed on the coronal aspect of the face by a pair of moderately large ones, which meet on the mesial line, and descend obliquely behind the nasals; and these by two larger pentagonal ones, lying one before the other and covering the crown of the head. A smaller rhomboidal scale interposes between them and the three small irregular ones that represent the upper eyelid (fig. 2). In like manner the margin of the mandible is protected by four large scales (fig. 2), sufficiently convex to form the third or fourth part of the circumference of a cylinder; and a single scale caps the symphysis, corresponding to the premaxillary one. The entire margin of the mandible shuts a little within the upper jaw. In the form of the scales between the limbs of the maxillary, represented in figure 4, *Anniella* differs from *Soridia*. Succeeding the symphysial scale, and between the marginal rows above mentioned, there is first a pair of moderately large scales which meet on the mesial line, then a row of four, and after that one of six. In *Soridia* a single pentagonal scale occupies the mesial space next the symphysial one, and the succeeding ones also differ from those of *Anniella*. The nasal scales of *Soridia* moreover have a small one beneath them on the edge of the jaw.

In *Soridia* there is a short, tapering, slender process on each side of the anus, representing limbs, and clothed by five rows of scales. These rudimentary limbs do not exist in *Anniella*, the anus being as represented in figure 5.

The *teeth* in both genera are alike, being small, subulate, slender, but not very acute, and disposed very distantly in a single series on the margin of the jaws.

Colour.—The general character of the stripes is very like that of *Soridia*, though the pattern differs. The ground-tint is silvery, with a conspicuous hair-brown line down each side of the back, in some places doubled; and very fine, faint, longitudinal zigzag lines, corresponding in number to the rows of scales, exist throughout the back, sides, and belly, the mesial one on the back being more evident than the others. A peculiar character is given to the surface by the closeness and smoothness of the scales; their disks are sub-pentagonal, and the zigzag lines of colour correspond to the meeting of the adjoining rows.

Dimensions.

Length from premaxillary to anus . . . . 4·5 inches.
Length from anus to tip of tail . . . . 2·8
Total length . . . . 7·3

The following Reptiles and Batrachians, named according to the Catalogue of the British Museum, were also collected on the voyage.

CANTON.

Hemidactylus frænatus . . . . by Dr. Seemann.
Gecko verus . . . . „
Rana rugulosa . . . . „
  „ *nova species* . . . . „

GALLAPAGOS.

Oreocephalus cristatus . . . . by Dr. Goodridge.
Leiocephalus Grayi . . . . „

ISLAND OF GORGONA, COAST OF PERU.

Phyllodactylus tuberculatus . . . . by Dr. Goodridge.
Gecko, near *Naultina* . . . . „

CALIFORNIA.

Hemidactylus Peruvianus . . . . by Dr. Goodridge.
Leiodera Chilense . . . . „

WESTERN MEXICO.

Crocodilus Americanus . . . . by Dr. Goodridge.
Goniodactylus . . . . by Dr. Seemann.
Tropidurus cyclurus . . . . „

MAZATLAN.

Rana, nova species . . . . by Dr. Goodridge.

Grammatophora and *Emys scabra*, from some locality not marked. Likewise upwards of fifty species of Snakes,—*Crotalus, Dendrophis, Boa, Pelamis*, etc., from various places, but chiefly from Western Mexico.

FISH.

A considerable collection of Fish was made by Drs. Seemann and Goodridge. The specimens put up at Canton, by the former of these gentlemen, are numerous; and, though of species already more or less perfectly described, are valuable for rectifying some points in the ichthyology of the China Seas. Professor Kaup having been lately employed in drawing up a catalogue of the *Plectognathi* in the British Museum, I took the opportunity of submitting that part of the collection to his inspection, and have adopted his names to the species described and figured in this work. He has divided the Linnæan genus *Tetrodon* into many groups, which he characterizes primarily by the form of the nostrils.

ANCHISOMUS GEOMETRICUS (Kaup).

(Anchisomus geometricus, Kaup, Cat. Br. Mus. Fish. ined.)

(Plate XXX.)

Of this Fish two specimens were obtained by Dr. Goodridge at the Gallapagos Islands, one of them considerably larger than the other.

Description.

Form oblong, with a moderately prominent belly, not largely distensible. Transverse diameter of the body greatest immediately behind the ventrals, and equal there to the vertical one. Snout distinguishable from the contour of the rest of the head, but not slender, the mouth being obtuse. *Nostrils* formed by a soft membrane arching over the nasal pit, and pierced on each side by a small round orifice, as is distinctly shown in figure 2. *Lips* fringed with soft, irregular papillæ. The *lateral line* commencing at the angle of the mouth and running with some slight undulations under the eye, then arching over the pectoral and continuing in a curve corresponding to that of the back and tail, reaches the caudal, keeping all the way along the upper part of the side; it is throughout a fine groove, and is crossed transversely behind the head by an equally fine line.

The *dermal spines*, represented of the natural size by figures 3 and 4, and magnified by figure 5, lie obliquely in the skin, with their points directed backwards and scarcely protruding through it. They are small, but rather larger on the belly than on the back, their form being the same in both situations. Their transverse thickish and crenulated base is shorter than the awl-shaped acute point. Anteriorly they terminate on the dorsal aspect between the fore part of the orbits, and posteriorly a little before the dorsal, while they descend on the sides to the level of the upper angle of the gill-opening. The circumference of the mouth, the cheeks, a broad ring round each eye, the flanks and tail, are smooth. On the ventral surface they extend from the chin to the vent.

Fins.—Br. 4; D. 7; A. 6; C. 9$\frac{1}{1}$; P. 17.

The *dorsal* commences over the vent, and is nearly the length of its base in advance of the *anal*. The first ray of both these fins is short, simple, and appressed to the second, and the last one is divided to the base. In the younger specimen the caudal ends in a regularly convex outline; but the older individual, which is the one that is figured, had sustained some injury in that member, as well as in the margin of the dorsal, which when perfect resembles the anal in its outline.

Both specimens have doubtless lost their original tints of colour by three years' maceration in spirits. In the young one the dorsal aspect is olive-green; in the old one, liver-brown. Both are rather thickly dotted with small, round, dark-brown spots, which are however scarcely so numerous on the young individual; on the other hand, this specimen shows more clearly various slender white lines traced on the dark ground, which have obtained for the species the designation of *geometricus* from Professor Kaup. Viewed on the dorsal aspect of the fish, the white lines form three or four concentric ovals (fig. 2), the outer ones being neither complete nor regular; the central one is in the middle of the back, and the outer one descends to the upper angle of the gill-cover. Several lines run between them, forming compartments, and they are continued undulating and anastomosing along the upper third of the tail. Three lines cross the face, one of them immediately in front of the nostrils. There are also many pores on the face. The ventral surface of the fish is white.

Length of larger specimen 9 inches.

Osteology.

The *vertebral column* consists of fewer joints in the *Plectognathi* than in most other orders of fishes which have a more perfectly ossified endo-skeleton. In this species there are 8 dorsal or abdominal vertebræ, and 10 caudals, of which latter the first one represents the sacral. On the lateral aspect the intercentral joints are all very apparent, but in several parts there are anchyloses of the neural and of the hæmal spines, with the interneural and interhæmal pieces respectively allowing of but little motion. The *centra* of the first three vertebræ are flattish on their sternal aspects, and partially confluent there: the first one is moreover embraced by the rather large, flat zygapophyses of the exoccipital, one on each side. A *hæmal canal* of small calibre, formed by the deflection of the *parapophyses*, commences with the fourth dorsal, and continues through the succeeding vertebræ onwards along the caudals. In the five posterior dorsals, a thin, almost transparent plate or hæmal spine runs along each centrum, forming by the series a very acute ridge on the sternal aspect of that portion of the vertebral column. Short *zygapophyses*, distinctly developed on the fifth and succeeding vertebræ to the third caudal inclusive, embrace the parapophyses of the immediately following vertebra. There are no *ribs*.

On the dorsal aspect of the column, a boat-shaped cavity is formed by the divergence of the right and left laminæ of the first four neural spines; it is obtuse at the atlantal end, where it receives the top of the occipital spine, and tapers to an acute angle at its termination in the spine of the fourth vertebra.

Anterior to the dorsal fin, there are two *dermo?-neural spines* lying end to end along the tops of the 6th, 7th, and 8th dorsal spines and the 1st caudal one. They are partly confluent with, and are supported by three *interneurals*, bent for that purpose in a knee-form. The dorsal itself stands over the spines and interneurals of the 2nd and 3rd caudals, all of them coalescent at their tips. Towards the middle of the vertebral column the *prozygapophyses* have more lateral prominence than on the more anterior or posterior vertebræ.

Hæmal spines springing from the first, second, and third caudal, lie parallel to and close along their respective centra, so that their tips are received between the edges of the succeeding pair of parapophyses.

Y

The remaining hæmal spines of the tail diverge downwards from the axis of the column, and expand, four of them into lanceolate laminæ, and the last three of the series more broadly still, and coalescing like the opposing neural spines, for the better support of the caudal fin.

At the base of the hæmal spine of the first caudal a slightly concave articulating surface facing downwards is formed by the parapophysis of that vertebra, to which there is jointed a long and strong interhæmal or pelvic bone, that, together with four other smaller interhæmals, supports the anal fin. These five interhæmals are confluent at their tips, and partially in the course of their lengths.

Skull.—In this species the coronal aspect of the cranium, formed by the frontal, parietals, and super-occipital, is flat or very slightly convex, both longitudinally and transversely, the triangular pre-frontals and post-frontals being depressed below the coronal plane, and also concave on that aspect. The projecting processes of the skull on its coronal aspect are the points of the triangular pre-frontals and post-frontals, an acute flat inial point of each parietal overlapping the lateral edge of the super-occipital, the mastoids, the petrosals, a crest at the upper angle of the operculum*, and the occipital spine. Compared to the next species to be noticed (*angusticeps*), this *Anchisomus* has a short and broad skull, the breadth of the interorbital space being equal to the length of the frontal and parietals. The occipital spine is of moderate length, and the inial angles of the parietals are small, as contrasted with their development in several members of the family, in which they are greatly elongated and hang over the mastoid fossæ. The pre-frontals are also rather small, and, like the post-frontals, are confluent with the frontal. A single slightly convex or arched nasal unites with the frontal by a superficial gomphosis, in which two acute scale-like processes of the frontal interlock with three of the nasal, the median point of the latter being produced in the line of the sagittal suture as far back as the middle of the orbits. Somewhat triangular turbinals, stout and about one-third shorter than the nasal, are attached to the antinial shoulders of that bone, and diverge, leaving a space for the pedicles of the premaxillaries. Their broad ends are articulated to the maxillaries, and by a corner to the premaxillaries, between which bones there is anchylosis at several points. *Pre-operculum* large, having its surface strongly marked by stout radiating ribs. Point of the operculum embraced by a deep notch of the nearly square and small *sub-operculum*. A slender bone, like a branchiostegal, articulating on the medial side of the hypotympanic with the mandible, and lying concealed behind the pre-operculum is most probably the *inter-operculum*, there being no other part of the skull to which I can assign that name. The antinial point of the pre-operculum interposed between this bone and the epitympanic also reaches the mandibular joint. In the hyoidean arch the *epihyals* are greatly developed, exceeding the coracoids in breadth, and being far larger than the cerato-hyals, which give attachment to the four branchiostegals. The upper point of the coracoid is crossed by the scapula, and projects behind it ; and the *epicoracoid*, instead of the usual styloid form that it possesses in fishes, has an acutely elliptical or broadly lanceolate lamina, which descends over the muscles of the body under the skin behind the pectoral fin. This bone has the same form in *Anchisomus angusticeps*, and its unusual expansion might be attributed to a necessity for its supplying the function of inflation in the absence of ribs; but in four other species of *Tetrodon*, of which three are known to be inflatile and of which Haslar Museum has complete skeletons, the epicoracoid is as usual slender and styliform. The proximal end of the epicoracoid, which is a separate slender bone joined splintwise to the dilated blade, is attached to the mesial side of the coracoid, just opposite to where it is crossed on its lateral surface by the scapula.

Pharyngeal teeth, in this species and *angusticeps*, small, crowded, slender, cylindrical, and obtuse.

* In the skull of a large *Tetrodon*, species and proper subgeneric group unknown, this crest is developed on the epitympanic ; but in the *Anchisomus geometricus* and *A. angusticeps* it is a projection of the posterior rib of the operculum, and if it has any connection with the epitympanic, there is such a confluence of the bones that their limits cannot be traced.

ANCHISOMUS ANGUSTICEPS (Kaup).

Tetrodon angusticeps, Jenyns, Zool. of the Voy. of the Beagle, p. 154, pl. xxviii.; Ann. 1842.

This member of the genus is wholly destitute of spiny armour, and is well described by the Rev. Leonard Jenyns in the work above cited, where there is also a figure. Dr. Goodridge obtained a specimen in the Gallapagos Archipelago, which has reached us in a very good condition.

Description.

The nostrils are minute openings, almost invisible to the naked eye, one on each side (lateral and mesial) of a loose or baggy flap, which is attached to the anterior margin of a small pit. When the flap reclines backwards it covers the tender blanched lining of the pit. It is probable that water flowing through the flap by its opposing orifices may erect it, and uncover the etiolated spot of membrane upon which it falls when relaxed. The olfactory nerve can be traced to this membrane and the base of the flap.

Mr. Darwin describes the living fish as dull green above, with the bases of the dorsals and pectorals black. These colours have perished in our specimen, which, like the one examined by Mr. Jenyns, is of a pale and tarnished wood-brown generally, and white beneath the tail. The skin is harsh to the finger when drawn over it in an atlantal direction, and under a lens is seen to be divided into minute, slightly elevated, flattish areas of very irregular form, but mostly tending to the orbicular. Mr. Jenyns notices two small, skinny appendages on the back, and in our individual there are five such processes, nearly in the same situation, but not symmetrically disposed, and seemingly the effects of the attacks of some parasite.

The *lateral line* commencing at the side of the mouth passes over the cheek to the posterior corner of the orbit and onwards to the side of the nape, where it makes an angle, beyond which it continues its course along the upper third of the side, undulating as it goes. Another line, equally fine and shallow, crosses the occiput, cuts the other a little behind the eye, and descends to the throat just before the humeral chain of bones.

Fins.—Br. 4; D. 1/8; A. 1/6; C. $9\frac{1}{1}$; P. 17.

The *dorsal* commences with a short acute jointless spine, having a branching ray behind it, so minute as to be visible only in the skeleton and with the aid of the lens. A similar spine, but with a joint or two at its tip, begins the anal, and it is not followed by the minute branching one. The last ray of both these fins is not divided to the base, as in the preceding species.

All the fins have transparent membranes more delicate than is usual in the group; the caudal having, however, strong opake rays. The anterior pair of rays in the dorsal and anal are abbreviated. The dorsal is situated over the vent more than its own basal breadth before the anal.

Length of specimen 15 inches.

Osteology.

In this fish the general structure of the skeleton nearly resembles that of *Anchisomus geometricus.* The number of vertebræ is the same. Two longitudinal dermo-neural bones overlie the tips of the last three neural spines of the dorsals, and the dorsal fin stands over the spines of the 1st and 2nd caudals. Connected with these five vertebræ there are nine interneurals, the less extensive ossification rendering their original number more easily determined than in that of *A. geometricus.* The anal is supported by five interhæmals, including the anterior strong one which articulates with the parapophyses of the first caudal. Excepting that anterior one, they have connection with the vertebræ only through membrane, not being intercalated with the hæmal spines which in all the caudal vertebræ lie appressed to the centra. Prozygapophyses, as in *geometricus,* most prominent under the dorsal and two dermo-neural spines which precede it.

Y 2

Cranium much more narrow on its coronal aspect than that of *geometricus*, the interorbital space being only about one-third of the length of the frontal and parietal.   The lateral edges of these bones, with the pre-frontals and post-frontals, are raised so as to make a deep channel along the crown, a sharp line marking the sagittal suture.   The inial points of the parietals are more developed than in the preceding species, and seemingly (judging from the course of the striæ of the bones) overlie a prominent lateral angle of the super-occipital, making together a truncated process.   There are however no remains of a suture marking the limits of the super-occipital.   Obtuse processes are also furnished by the mastoids and petrosals.   The posterior rib of the operculum is a thin plate standing at right angles with the disk of the bone, and having a more prominent shoulder at its upper end.   Anterior to this, and completely separated from it, there is an irregular process of the epitympanic, which does not exist separately, at least in *geometricus*.   The nasal is long, slender, and straight, and the turbinals, which are stouter, broader, and as long, are attached to its antinial end.   They are broad at their extremities, where they articulate with the maxillaries and pre-maxillaries, and pointed at the opposite end.   The anterior limb of the preoperculum is elongated, to match the length of the nasal and turbinals, and there is no bony septum between the orbits, in which this species differs from *geometricus*.   *Epicoracoid* of two pieces, the distal one dilating into a thin, broadly lanceolate or elliptical plate.

ANCHISOMUS MULTISTRIATUS (Kaup).

(*Anchisomus multistriatus*, Kaup, Cat. Br. Mus. Fish. ined.)

(Plate XXIV.)

Figs. 1 and 2, half the linear dimensions ; fig. 3, natural size.

The only specimen of this Fish in the collection is a dried one ; the structure of the nostrils and tints of colour are therefore lost, but the markings shown in the plate are discernible.   It is an inhabitant of Southern Polynesia.

Description.

Form much like that of *Anch. geometricus*.   In one nostril the shrivelled, perforated membrane remains ; but in the other, which happens to be on the side selected by the artist for representation, it is destroyed.   Dorsal in advance of the anus, its last ray being over the commencement of that orifice.   Some little allowance may be made for displacement in stuffing the skin, but there is no apparent distortion in the specimen.

Fins.—D. 11 ; A. 9 ; C. 9½ ; P. 16.

The first ray of the dorsal is very short, and incumbent on the second.   In the anal the corresponding ray is also incumbent, but is not so short.   Caudal nearly even.   Spines are represented by figure 3, with the two parts of the base meeting at an angle, from which there springs the subulate point, in general shorter than either of the basal portions.   A detached spine is in fact a caltrops with three prongs, and when in its natural position in the skin the short point inclines towards the tail and scarcely penetrates the integument.   A single spine stands on the mesial line immediately before the nostrils ; six in two rows occupy the top of the head ; on the occiput there are three in one transverse row.   From thence their anterior limit curves down behind the temples and cheek to opposite the angle of the mouth, where the first spine stands under the nostrils.   The snout, cheeks, chin, and circumference of the orbit, are naked.   Posteriorly the spines end before the dorsal, but not by an even line, and on the belly they extend to the border of the anus.   They are rather widely apart on the back, but are more numerous and have longer points on the ventral aspect.   The belly is finely plicated, as if capable of considerable distension.

The markings are a series of pale or whitish narrow loops, extending obliquely forwards on the sides and cheeks, the areas being dark. On the posterior part of the sides and on the tail the loops are crenulated or beaded and interrupted, with the interstices flecked by short bars. On the middle of the back there is a series of concentric narrow and acute longitudinal ellipses. The lines on the face are also longitudinal, but on the sides of the tail the pale lines form reticulations insulating some roundish blotches. The dorsal and anal are clouded.

Length 16·5 inches.

ANCHISOMUS RETICULARIS (Kaup).

(*Anchisomus reticularis*, Kaup, Cat. Br. Mus. Fish. ined.)

(Plate XXXI.)

Figs. 4, 5, nat. size.

This species inhabits the western coasts of the South Atlantic, extending northwards to Jamaica, from whence we have an overstuffed dried specimen. The figure is a representation of a rather smaller individual in better preservation.

Description.

Form more elongated than some others of the genus above described. The nose especially is lengthened considerably before the nostrils, which is chiefly due, as far as can be judged from the undissected specimen, to the long and strong turbinals stilting out the upper jaw. These bones are represented in figure 5 shining through the dried integument on each side of the premaxillary pedicles. There is a prominent, oblique, ridge-like projection of the prefrontal, shown in figures 4 and 5 as existing between the eye and the minute membrane of the nostril, which latter organ is situated midway between the occiput and the upper-jaw teeth. The face is also narrower transversely than is usual in the *Tetrodons*. *Gill-opening* small, on a level with the upper half of the pectoral. Dorsal small, situated over the interspace between the vent and anal: the latter fin being narrower than even the dorsal. Caudal pretty large, and even at the end.

Fins.—D. 8; A. 7; C. 9¼; P. 14.

The first ray of the dorsal and anal is short, and applied closely to the base of the succeeding one.

No trace of a *lateral line* was detected in the specimens, but many pores on the face and cheek are visible.

Dermal spines small (fig. 6), with a stellate base of three, four, or more generally five short rays, and a still shorter subulate point rising from the centre. The circumference of the mouth, the chin, a rather narrow border to the eye, axilla of the pectoral, a stripe in the middle of the side commencing over the anus and running backwards, and the entire tail posterior to the vent, are smooth. The rest of the surface is crowded with spines, most densely on the nape, and nearly as much so on the belly, where they approach the vent in one direction, and the angles of the mandible on the other. A narrow smooth space precedes the dorsal; and on the coronal aspect of the face the spines are more distant and in two alternate series, that go forward to the points of the premaxillary pedicles.

The *colour* of the specimens is lost, but on the dorsal aspect many small polygonal areas are traced on a dark ground by white reticulations. A double row of larger and darker spots, not regularly disposed, can be traced from the corner of the mouth to the gill-opening, and from thence along the flanks and the upper part of the tail to the caudal. The belly is white, and the fins pale, or perhaps colourless.

Length of the subject of the figure, 7·5 inches. The other specimen is a little larger.

The species of this genus mentioned in Dr. Kaup's list are *Anchis. Spengleri*, *angusticeps*, *multistriatus*, *reticularis*, *scalaris*, *geometricus*, and *turgidus*. *Anchisomus*, *Gastrophysus*, *Cheilich thys*, and *Leiosomus*, form a group of *Tetrodontidæ*, in which the "nasal cavity is small and flat, with two nostrils."

PRILONOTUS *vel* ANCHISOMUS CAUDACINCTUS.

(Plate XXX.)

Figs. 1 and 2, nat. size; fig. 3, magnified.

The single specimen now under consideration, having been submitted to Professor Kaup's inspection, along with other *Plectognathi* belonging to the Museum of Haslar Hospital, was returned labelled "*Prilonotus nova species;*" and in the Professor's manuscript catalogue, the distinctive character given of the genus is "nasal cavity flat, without any nostrils." *Prilonotus* is a name invented by Müller, and is mentioned by him in his 'Fortsetzung der Myxinoden,' and in the 'Archiv für Naturgeschichte für 1841,' but I have not found his detailed account of the characters. Dr. Kaup enumerates the following species:—*Pril. rostratus*, Lin. (*margaritatus*, Rüppell, *Solandri*, Richardson), *millepunctatus*, *occipitalis*, *oculifer*, *insignitus*, *cæruleo-punctatus*, and *pictus*. In *Pril. occipitalis* and *oculifer* the olfactory apparatus is a small oval saucer with slightly elevated edges, and a pale disk of a different texture from the surrounding integument, and no perforations or nostrils. On taking our specimen of *caudacinctus* out of the spirit, no nasal pit or nostril could be perceived; but after maceration for an hour or two in water, a minute spot on each side of the nose became just visible to the naked eye, and when examined through a lens was seen to be a depression filled by a soft, swelling membrane, in which no apertures could be discovered. Perhaps this membranous covering was merely the puffing up of the usually adherent lining of the nasal saucer of *Prilonotus*, detached by maceration and handling.

The locality where the specimen was captured has not been noted.

Description.

Form of the head and body not much unlike that of *Anchisomus reticularis*, but the face is still more elongated anterior to the nostrils, while the tail on the contrary is thicker and shorter, and the summit of the back is over the gill-opening. A view of the back of the fish, given by figure 2, shows that the greatest transverse diameter is at the same part, and that the outline of that aspect of the fish is broadly fusiform. Minute, almost microscopic nostrils exist a short way before the eye, each consisting of a depression covered by a soft membrane. Were the membrane perforated by an aperture on each side, the nostrils would be exactly like those of the *Anchisomi*. The mucous secretion which exudes when the fish is placed in strong spirit readily hides these organs, and they become imperceptible also when the skin is allowed to dry, but they become apparent after the specimen has been soaked in fresh water. There are several microscopical pores scattered on the sides of the face. Orbits prominent, with a wide concavity between them. Lips fringed with slender papillæ.

Fins.—D. 10; A. 9; C. 9¼; P. 15.

The pectoral is broad but of moderate size, and is truncated or slightly concave on the edge, with corners rounded off. The dorsal is placed anterior to the vent, which, contrary to its position in the preceding species, is situated at the base of the first anal ray. Both fins are rounded, and have a short first ray. Caudal even at the end.

Lateral line imperceptible. Skin minutely rugous and on the belly plicated, and consequently distensible. At first sight no spines are visible on the back; but on allowing the specimen to dry for some time, minute ones, having a slender curved base, and a somewhat shorter subulate point, are observed to be scattered over the greatest part of the body anterior to the anus. They are more numerous on the occiput and nape, and fewer on the face, but they surround the eyes and leave only a narrow naked space at the mouth. The smooth axillæ of the pectorals are also restricted. On the back the spines run backwards to the dorsal, and one or two even pass that fin. They do not extend quite so far back on the flanks. The tail is destitute of them. On the belly each spine is lodged in a canal with a round open mouth, as represented magnified by a common eye-glass at figure 3, where the crescentic base of the spine is shown through the integument, and, as it usually is, enveloped in a sheath, and consequently thicker than it is in reality.

The *colours* are altered, and no doubt partly obliterated by long maceration in spirits. At present the specimen is light brown above and whitish below. Some pale dots on the posterior part of the back, run into rows near the caudal. The under half of the tail and the chin are crossed transversely by chestnut-coloured bars; three of the same hue radiate with a curve from the eye towards the nostrils, and four from the eye curve over the temples. There are also some markings of a similar tint on the outer rays of the caudal, and a single purplish line. The pectorals, dorsal, and anal are colourless, but a slight trace of a dark spot exists on the back below the first dorsal rays.

Length 3 inches.

TETRODON VIRGATUS (Richardson).

Tetrodon virgatus, Richardson, Zoology of Voy. of the Erebus and Terror, p. 62, pl. xxxix. f. 8 and 9.

(Plate XXVIII.)

Figs. 6 and 7, nat. size; fig. 8, magnified.

This species belongs to Kaup's restricted genus of *Tetrodon* proper, the members of which do not possess nasal orifices, but have merely a pair of imperforate barbels in the usual site of each nostril. The fish is an inhabitant of the Indian Ocean, and there are specimens in Haslar Museum both from Torres Straits and Port Jackson. A figure of the species having been given in a preceding work, it was repeated in this through a mistake in the delivery of specimens to the artist; but I am enabled thereby to correct an error committed in my first description of the fish, wherein the nasal barbels are stated to be tubular.

Description.

Form elongato-ovoid, inflatible. Dorsal in advance of the anus, between which and the first anal ray there is a fold of skin. Nasal barbels two on each side, united at the base, flat, strap-shaped, imperforate and not tubular, the integument of one flat side of each barbel being closely connected with those of the other by cellular substance; surface rugous, forming a multitude of shallow pits*. The surface of the skin

* An accidental notch in the summit of one of the barbels in a Port Jackson specimen caused me to describe the barbels as tubular in the Zoology of the Voyage of the Erebus and Terror. Professor Kaup's character of the genus led to the examination of the specimens again, and the discovery of my error: on dissecting the barbels, no exterior opening was detected, but their base was found to be bound down to the nasal space before the orbit by a column of fibrous matter, in the centre of which lay the olfactory nerve, of moderate size.

of the body and tail is divided into square segments, which are visible by aid of a common lens, and appear to be generally of one size.   On the belly there are longitudinal folds, with many round or oval pits, each lodging a spine, which is clothed with white membrane nearly to the summit.   This membrane is represented in figure 8, and gives much apparent thickness to the spine, which is in reality very slender and bristle-shaped.   The base of the spine is very short and consists of two arms, that together form a shallow crescent.   The lips, a narrow ring round the eye, circumferences of the fins, edge of the gill-opening, and greater part of the tail are unarmed; the rest of the integument of the head and body is densely studded with the slender erectile spines, most of them, though rigid, nearly as fine as a human hair.

   Fins.—Br. 4; D. 9; A. 8; C. 9$\frac{1}{1}$; P. 15.

   The dorsal is in advance of the vent, which is separated from the base of the first anal ray by a fold of skin.   Caudal slightly rounded at the end.

   Colour. — The upper parts are blackish-brown, and the lower ones dusky-white, but both are most likely altered by long maceration in spirits.   These ground-tints are traversed longitudinally by black lines, those on the dorsal and ventral aspects beginning near the mouth, and each meeting its fellow at an acute angle behind the dorsal and anal fins respectively.   On the sides the lines form concentric parabolic curves round the pectoral, the open branches of the parabola running on to the caudal.

   Length 4$\frac{1}{2}$ to 5$\frac{1}{2}$ inches.

Osteology.

   In this species the boat-shaped cavity formed by the anterior neural spines extends into the fore part of the 5th vertebra.   The 6th, 7th, and 8th spines support the longitudinal dermo-neurals, and the dorsal is seated over the first and second caudals.   There are, as in the preceding species, 8 dorsals and 10 caudals.   Six of the hæmal spines next the caudal fin diverge from the axis of the vertebral column.

   The cranium has a general resemblance, in comparative breadth, etc., to that of *geometricus*, but is more convex on its coronal surface, and its occipital spine is stronger and longer.   There are also some differences in the prominent angles of the skull, but the skeleton is too much injured to admit of exact comparison.   The *epicoracoid* consists of two pieces, and is long, slenderly subulate, and moderately curved.

PLATESSA STELLATA.

(*Pleuronectes stellatus*, Pallas, Nov. Act. Petrop. i. p. 347, An. 1783; Tilesius, Nov. Act. Petrop. i. p. 387.
    tab.   , f. 1. An. 1787; Mémoires de l'Acad. de Pétersb. iii. p. 248. t. x. f. 1.   Richardson, Fauna
    Boreali-Americana, Part 3rd, Fishes, p. 257, An. 1836.)

(Plate XXXII.)

Figs. 1–3 (the figures reversed).

   On the 16th of July, 1821, we took several flounders in nets set off the mouth of the Coppermine River, which I had little difficulty in recognizing as the "stellated flounder" of the Sea of Kamtchatka and Beering's Straits, the "Cambala" of the Russians, and the "Tanticu" of the inhabitants of the Kurile Islands. At that early period of Arctic discovery, the taking of a fish common to Beering's Sea and Hearn's Icy Ocean, afterwards named "Coronation Gulf," was considered to be an event of good augury.   The melancholy and disastrous journey from the coast, which commenced about six weeks afterwards, necessarily occasioned the abandonment of all our specimens of natural history; but a description of the recent fish recorded on the spot was published in the appendix to Sir

John Franklin's Journey in 1822, and in the 'Fauna Boreali-Americana' fourteen years later. Mr. Collie, surgeon of the Blossom, under Captain Beechey's command, in 1827, observed the species in Awatscha Bay; and Dr. Rae has recently furnished me with specimens captured in the locality where I first observed it, from one of which Plate XXXII. was executed. Unfortunately I omitted to caution the artist to reverse his drawing on the stone, and it thence becomes necessary to remark that the dark side of the fish is the left one, and not the right one as in the figure.

Description.

Rays.—Br. 7–7; D. 55; A. 41; C. 15½; P. 10; V. 6.

The *dorsal* commences over the anterior third of the upper orbit, and ends about a ninth-part of the entire length of the fish before the base of the caudal. It attains its greatest height at the 30th and 31st rays, which stand posterior to the middle of the fish. The anal is similar in shape and depth to the dorsal, except that it has not the same anterior extension, its commencement corresponding with the 15th or 16th dorsal ray. No difference exists between the pectorals in their size or in the number of their rays, but the right one is more delicate and colourless, like the rest of the under side. Both ventrals are also alike in size and form. A small pelvic spine projects forwards at the beginning of the anal.

The outline of the fish, including the dorsal and anal, but excluding the tail beyond these fins, shows the usual rhombic form of the genus.

Teeth in a single series, cutting, but individually narrow at the crown, though not acute. There are 15 in each premaxillary, and as many in each limb of the mandible. The jaws are flatter on the right side, but the maxillaries do not differ from one another in size.

On the left or dark side, the head, body, and tail are generally studded by small scales bristling with spinous points of unequal length (Fig. 3). These scales are more crowded on the tract from the upper eye to the parietal crest, on the cheeks, throat, and tail. They are larger, and form a continuous line along the base of the dorsal and anal, and onwards on the margins of the tail to the caudal. A few of smaller size encroach on the basal half of this fin, but none exist on the rays or membranes of the other fins. On the pale or right side of the fish the muricated scales are fewer, and for the most part smaller. The rows are pretty continuous along the bases of the dorsal and anal fins, and there is a row on each side of the spine; otherwise they are distributed as shown in fig. 2, considerable spaces of this side of the fish posteriorly being naked.

Lateral line beginning at the occiput, arching over the pectoral, and then running along the centra of the vertebræ, and between the middle rays of the caudal to the extremity of that fin. On the right side the curve it makes over the pectoral undulates, and is less bold.

Colour of the recent fish on the left side liver-brown without spots, white on the other side; the vertical fins barred in the direction of their rays with black.

Dimensions.

Length including the caudal fin	.	.	.	.	.	.	.	.	.	8·6 inches.
Length excluding the caudal fin	.	.	.	.	.	.	.	.	.	7·0
Height of body excluding the fins	.	.	.	.	.	.	.	.	.	4·0
Height of body including the fins	.	.	.	.	.	.	.	.	.	6·0

Osteology.

Vertebræ 35; the strong pelvic bone supported by the hæmal spine of the 12th. Anteriorly and posteriorly, the interneurals and interhæmals correspond in number with their respective neural and hæmal spines, but under the higher thirds of the dorsal and anal fins they are doubled.

z

PLATESSA GLACIALIS.

(*Pleuronectes glacialis*, Pallas, Reise durch versch. Prov. Russlands, 1772–73, p. 706 ?   Richardson, Fauna
Boreali-Amer. Fishes, p. 258.)

(Plate XXXII.)

Fig. the head only reversed.

Pallas, in his account of his journeys through various provinces of Russia, describes in the following terms a *Pleuronectes* which inhabits the Icy Sea :—" *Pleuronectes glacialis*, dodrantalis, facies
Flesi.   Oculi a latere dextro fusco, subaspero ; latus album læve.   Spinæ nullæ, nec ad pinnas, neque
in linea laterali.   Tractus capitis, pone oculos prominulos, scaber, sed non in tubercula divisus.   Radii
medii pinnæ dorsi anique a latere fusco quasi spinulis minutissimis hispidati.   Radii pinnæ dorsi 56,
ani 39.   Frequens in oris arenosis Oceani glacialis.''   On the 5th of August, 1821, we captured in
Bathurst's Inlet a flounder, of which I took a short description ; but the fish itself being required for
food to the party, no specimen was retained.   Finding, after my return to England, that my description, as far as it went, agreed so closely with the passage quoted above, I introduced it into the ' Fauna
Boreali-Americana,' adopting the specific name given by Pallas.   Dr. Rae has since then sent me two
specimens from the same sea, and they present all the characters enumerated by Pallas, with the exception of the roughness of the central rays of the anal and dorsal fins.   As this however may be
only a sexual peculiarity, or Pallas may have intended to allude merely to the skin at the bases of
the rays, I still retain the name for the species, in the belief that the fish is an inhabitant of the
Siberian as well as the Arctic American seas.

Description.

Form.—The circumscription of this fish, with the fins spread out, is less acutely rhombic than that of
Platessa glacialis, the deeper parts of the dorsal and anal being rather rounded than angular.

Rays.—Br. 7–7 ; D. 58 ; A. 43 ; P. 11 ; C. $14\frac{2}{2}$; V. 6.

The dorsal commences at the fore part of the orbit.   That fin and the anal attain their greatest depth
about the middle of the total length of the fish, or at the 32–34th rays of the former, and the 15–18th rays
of the latter.   The point of the pelvic bone ends anteriorly in a small spinous point, which protrudes
through the skin at the base of the first anal ray.

Lateral line quite straight from the suprascapular to the extremity of the caudal, coincident with the
centra of the dorsal vertebræ, and unarmed.

Scales small, circular, smooth, deeply imbedded in mucous skin, and having their disks marked by fine
grooves radiating from a spot near their posterior border.   They are not tiled, touching each other only in
a few places, and they present no roughness to the finger drawn over them in any direction, except along
the base of the dorsal, where two or three rows of the scales have a fringe of short spiny teeth on their
posterior borders, so minute as to be scarcely perceptible to the unassisted eye.   These minute spines do
not become visible even by the aid of a lens, anterior to the 20th dorsal ray, and cease again opposite the 9
last rays.   Similar rows of spiny scales commence on the ventral margin of the fish with the 12th anal ray,
and end with the 37th ; they are confined to the right or coloured side of the fish.   On the tail, the
ordinary unarmed scales are more crowded, and they extend over the cheeks and gill-covers, none however
being visible on the limb of the preoperculum.   The scales on the left uncoloured side of the body are
more delicate, still more deeply imbedded in the skin, and quite unarmed even at the dorsal and ventral
borders.

Between the orbits there is a narrow, elevated, smooth ridge. The parietal and suprascapular space is divided into elevated granulated surfaces by smooth furrows as represented by figure 4.

Orifice of the *mouth* more restricted on the right or dark side; the right premaxillary having only 5 teeth, while the left one carries eighteen, and the dark limb of the mandible is in like manner armed with four teeth, while the pale one has 21. These teeth form a very even close series, their crowns being as thick as their stems. Towards the articulations of the mandible the crowns are somewhat chisel-shaped, but they are blunt nearer the symphysis, and so also on the premaxillaries*.

Dimensions.

Length exclusive of the caudal	7·5 inches.
Greatest height of body	3·6
Length including caudal	8·8
Height or breadth including dorsal and anal	5·0

Osteology.

Vertebræ 40, the stout pelvic bone supported by the hæmal spine of the 13th. As in the preceding species, so in this, there are a pair of interneurals or interhæmals between each two corresponding spines in the middle of the dorsal and anal, but only one for each spine towards either extremity of these fins.

SALMO CONSUETUS (Richardson).

(Plate XXXIII.)

Fig. 1 and 2, half the natural size.

The Salmon which we introduce under the above name is an anadromous species, which ascends the river Yukon to the falls above the confluence of the Porcupine, and perhaps even higher. It is one of those so nearly resembling the *salar* that when seen without the means of direct comparison it appears to be almost the same, and even when the specimens are contrasted the distinctive characters are by no means striking. The example which has reached us, is the stuffed skin, most probably of a male, if we may consider the curvature of the premaxillaries as characteristic of that sex in the spawning season. The length of the upper jaw, which passes beyond the mandible, and the large gape of the mouth, some differences in the outline of the gill-flap, and consequently in the shape of its component bones, together with the comparatively smaller scales, distinguish it from the well-known and much prized salmon of our markets. It does not attain the size of the latter, and the dimensions of our specimen are merely those of a well-grown sea-trout. I have been unable to identify it with any of the species described by Pallas in his ' Zoographia Rossica.'

Description.

The general form is that of a common salmon out of season, when the head is disproportionately large and the jaws distorted. In a well-formed British *Salmo salar* in prime condition, the head appears small as compared to the body either in thickness or height, its length to the distal margin of the gill-cover being

---

* Through hasty and inaccurate observation on the sea-coast, in 1821, I erroneously stated the teeth to be brush-like.

contained five times and three-quarters in the total length of the fish, caudal included ; the gape of the mouth is also small, the tip of the maxillary scarcely passing the centre of the eye ; and the acute mandible equals or is even a little longer than the upper jaw, which is somewhat blunter. In *consuetus* the length of the head is contained only four times and two-thirds in the total length of the fish, or four times in the distance between the tip of the snout and base of the caudal fin. The premaxillary symphysis is acute, and hooks over the front of the blunter end of the mandible, passing it by about half an inch, and the tip of the maxillary goes some distance behind the orbit. The distal edge of the gill-flap, which in *salar* is parabolic, in *consuetus* is two sides of a rectangle, with the corner of the sub-operculum slightly rounded off. From this comparison of the gill-flaps, the branchiostegal rays are excluded ; but when they are extended in *consuetus*, they form, in conjunction with the operculum and sub-operculum, a segment of a circle, while in *salar* the segment would be the end of an obtuse oval.

Teeth.—Five or six recurved teeth, most of them rather large, stand on each premaxillary ; they are followed by an unarmed hollow which receives the tip of the lower jaw, and then by the single series of small, acute maxillary teeth, about 20 in number, four or five of the anterior ones being larger and more remote than the posterior ones. The palatine teeth stand also in a single series on each side, and are rather smaller and less numerous. Two small recurved teeth arm the front of the vomer immediately behind the edge of the velum, and they are succeeded by two still smaller ones, one before the other on the mesial line. Two parallel rows of six teeth each, equal in size to the larger maxillary ones, form the armature of the tongue.

Scales.—In delicacy, lustre, and the ease by which they are detached, the scales resemble those of the common salmon ; but they are of smaller size, in the proportion of 12 or 13 to 10, when equal portions of integument with the scales *in situ* are compared. Detached scales of the two species differ to a still greater degree in area. The scales on the lateral line of *consuetus*, when reckoned to the extremity of the rectangular patch on the caudal, number 140.

Rays.—Br. 12–13 ; D. 11–0 ; A. 15 ; C. 17⅜ ; P. 14 or 15 ; V. 9–9 or 10–10.

The fin membranes are thicker than those of *salar*, and it was only by dissection of the dried specimen, after soaking, that I could ascertain the number of their rays. The artist, having mistaken the folds of membrane for rays, has introduced too many into his representation of the caudal fin. The last ventral ray is divided to the base, the posterior portion being slender and short.

Dimensions.

Length from tip of premaxillaries to lower corner of the caudal	23·5 inches.
Length of middle caudal rays from the termination of the scales on their base to their branching tips	1·6
Length of either the upper or lower caudal lobe from its base . . .	3·8
Depth of the fork or notch of the caudal fin when expanded	0·7
Length from tip of premaxillaries to the first ray of the ventrals . . .	12·0
Length from tip of premaxillaries to the first ray of the dorsal . . .	10·2
Length from tip of premaxillaries to the first ray of the anal . . .	15·9
Length from beginning of scaly integument on the nape to the dorsal fin . .	7·1
Length from the tip of the premaxillaries to distal edge of the gill-cover . .	5·2
Length from the same point to the edge of the scaly integument on the nape .	3·6
Length from the same point to the centre of the eye	2·4
Length of the maxillary on its basilar edge	1·8
Greatest width of its lanceolate tip	0·42
Distance of the nostrils from the tips of the premaxillaries	1·6

Of the salmons described by Pallas this fish appears to approach to *callaris* (Zoograph. p. 352), which frequents the Sea of Okhotsk. This is said to have the first ray of the pectorals and ventrals snow-white, but in our dried specimen these rays are darker than the others. Pallas, in his specific character, makes the mandible shorter than the snout, though in giving the details he describes it as longer. It possesses also many of the characters noted by the same author in his description of *S. Lycaodon* (p. 370); which frequents the same seas with *callaris*, and agrees tolerably with it in the number of its fin-rays; but without comparison of specimens or of good figures it would be unsafe to affirm that they are specifically the same.

———————

In the Straits of the Dolphin and Union, and off the mouth of the Coppermine River in the Arctic sea, we observed many small salmon, which in general aspect, the numbers of the fin-rays, etc. resembled the widely-spread *salar* still more nearly than the preceding; so much so indeed, that had I not perceived the gape of the mouth to be larger, and thought the scales to be rather smaller, than those of the well-known species, I should have considered them to be the same. From a sketch which I made on the spot, I find the outline of the gill-flap to be exactly like that of our English fish.

Description

(drawn up at the place of capture).

Rays.—Br. 12; D. 13; A. 11; V. 9; C. 23⅝; P. 15.

The first three dorsal rays short and incumbent on the fourth, and the last anal ray divided to the base. Caudal fin even at the end when extended.

Head one-fifth of the total length, caudal included. Jaws nearly equal, the lower one scarcely perceptibly advancing further than the acute snout. Maxillary passing considerably beyond the small eye.

A single series of conico-subulate very acute teeth arms the premaxillaries, maxillaries, mandible, and palatines. A double row occupies a small triangular space on the vomer, and there are two rows on the tongue not exceeding in size those on the jaws.

Dimensions.

Total length, caudal included	29·0 inches.
From the tip of the snout to distal margin of gill-flap	5·75 inches.
From the tip of the snout to ventrals	13·5
Length of central caudal rays	2·0
Girth of the thickest part of the body	16·0

The *colour* bright and silvery, like that of a salmon in good condition, with faint spots adjoining the lateral line above and below it. The flesh is of a bright pinkish-red.

SALMO DERMATINUS (Richardson).

(Plate XXXIII.)

Figs. 3 and 4, half nat. size; fig. 5, magnified.

This species, known locally to the fur-hunters on the Yukon by the name of "Red-fish," is very distinct from every variety of the *Salmo salar*, and cannot be confounded with any of the European anadromous salmons. It ascends the Yukon from Beering's Sea in vast multitudes, and is the object of native fisheries established at convenient places on that great river. The chief capture is made in weirs constructed between island and island on shallow expanses of the stream, and multitudes are speared in the eddies. The flesh and roe are dried for winter consumption, and the skins are tough enough to be employed in forming articles of clothing. This species is the "tleukh-ko" of the Kutchin, according to Mr. Murray of the Hudson's Bay Company, who procured the specimen.

Description.

Teeth.—None on the tongue; 7 or 8 on each premaxillary, of unequal sizes, one or two at the tip larger than any of the others, conical, acute, and recurved. A small space intervenes between the last premaxillary tooth and the maxillaries, but it has no reference to the extremity of the mandible. On the maxillary there are from 18 to 21 acute, straight, subulate teeth, the first being the stoutest, and the next in size being the third and fifth. Three small ones, one before the other, arm the vomer; and there are only two on each palatine, both of them as large as the first maxillary one. On each limb of the dilated, almost circular apex of the mandible there are 5 teeth, of which the anterior one is the largest of all, the next two equal the biggest of the premaxillary ones, and the fourth and fifth resemble the remaining 15 situated on the limb of the jaw, which correspond in size with the posterior maxillary teeth, and have a tendency to be alternately larger and smaller, but are not uniformly so. The edge of the velum is not so transverse as it is in most other salmon, but runs backwards on the lateral side of each palatine bone.

The end of the snout is obtuse, though less so than that of the mandible, which forms a roundish knob, considerably wider than the limbs of the jaw immediately behind it. The knob is inclined upwards, and, being nearly of the same length with the incurved snout, the two meet and prevent the closure of the mouth, as shown in the profile, figure 3. Figure 4 gives the end view reduced, like figure 3, to half the natural size. The gape of the mouth, and the tip of the long, slender maxillary, pass the orbits.

Scales.—The most striking distinctive character of this salmon is the thick muciferous epidermis in which the scales are firmly imbedded. The scales themselves are comparatively small, and when detached are seen to have a compressed oval form placed vertically: as they are however in no way tiled, and do not even touch each other, they seem larger when *in situ* than they actually are. A slightly enlarged view of a portion of the integument at figure 5 shows their arrangement. They number about 138 on the lateral line, including 7 or 8 rows of smaller ones covering the base of the caudal.

Rays.—Br. 14–15; D. 15; A. about 15; C. 17$\frac{10}{10}$; V. 9–9 or 10–10; P. 16.

The first three rays of the dorsal are graduated, simple, and closely applied to the fourth, which, with the succeeding ones, is largely branched. The same seems to be the case with the anal; but it was only by dissection that the numbers of the rays of either could be ascertained, and, from the hardness and toughness of the dried skin, this could not be satisfactorily accomplished with the anal without doing too much injury to the specimen. A rounded anterior tip is given to the anal in the figure; but the edge of this part of the fin is broken away, and the degree of rounding it originally had cannot be learnt from our solitary specimen.

Dimensions.

Length from tip of snout to the extremity of the lower caudal lobe . . .	32·4 inches.
Depth of the notch in the margin of the caudal	1·0
Length of the middle caudal ray beyond the scales which cover its base . .	1·8
Length of each caudal lobe, from the attachment of its rays . . .	5·0

Length from tip of the snout to the ventrals . . . . 17·2 inches.
Distance between the first rays of the pectorals and ventrals . . . . 9·8
Distance from the smooth skin on the hind head to the dorsal . . . . 11·0
Distance from the tip of the snout to the extreme edge of the gill-flap . . . 7·2
Distance from the tip of the snout to the centre of the eye . . . . 3·5
Length of the basilar margin of the maxillary . . . . 3·0
Greatest width of that bone, half an inch from its distal end . . . . 0·35
Length from end of the snout to the nostrils . . . . 2·2

Colour.—The cloudy patches represented in the figure exist on the specimen, and there is much dark marking on the belly. The fins also are dark, with the exception of the tips of the anal and ventrals, which are orange in the dried specimen. We have no means of knowing the tints of these and of the other parts in the fresh fish.

This is a very distinct species from the *Salmo Scouleri* of the 'Fauna Boreali-Americana' (Part Fishes, pl. xciii. p. 158), which has still smaller and very different scales. Pallas has omitted to mention the armature of the tongue in his account of the *Salmo proteus*, which in some of the characters he mentions agrees with this fish, though it does so still better with *Salmo Scouleri*. *S. purpuratus*, *S. sanguinolentus*, and *S. Lycaodon*, of the same naturalist, all frequenting the seas of Kamtchatka, are described by him as having toothed tongues. In some others of the same waters the tongue is not spoken of.

Haslar Hospital, June 21, 1854.

THE END OF THE VERTEBRALS.

ERRATA.

Page 20, line 23, *for* the ulna measures, *read* the tibia measures.

Page 33, line 14 from bottom, *for* fig. 6, *read* fig. 5 ; and in line 11 from bottom, *insert* Plate VIII. before fig. 1.

Page 45, last line, for *Third cervical*, read *Fourth cervical*.

Page 47, No. 116, for *fourth*, read *fifth*.

Page 49, line 9, *for* fifth, *read* third ; *for* third, *read* fourth ; and *for* fourth, *read* fifth. These several corrections imply some reformation in the comparisons with the homologous parts of the ox. The dimensions of the fourth and fifth cervicals of the latter animal are recorded on page 48.

Page 58, line 7 from bottom, *for* 16·5, *read* 18·1.

Page 84, line 12, *for* radius on its rotular, *read* radius on its thenal.

Page 147, line 8, *for* coracoid, *read* coranoid.

JOHN EDWARD TAYLOR, PRINTER,
LITTLE QUEEN STREET, LINCOLN'S INN FIELDS.

Plate XXV.

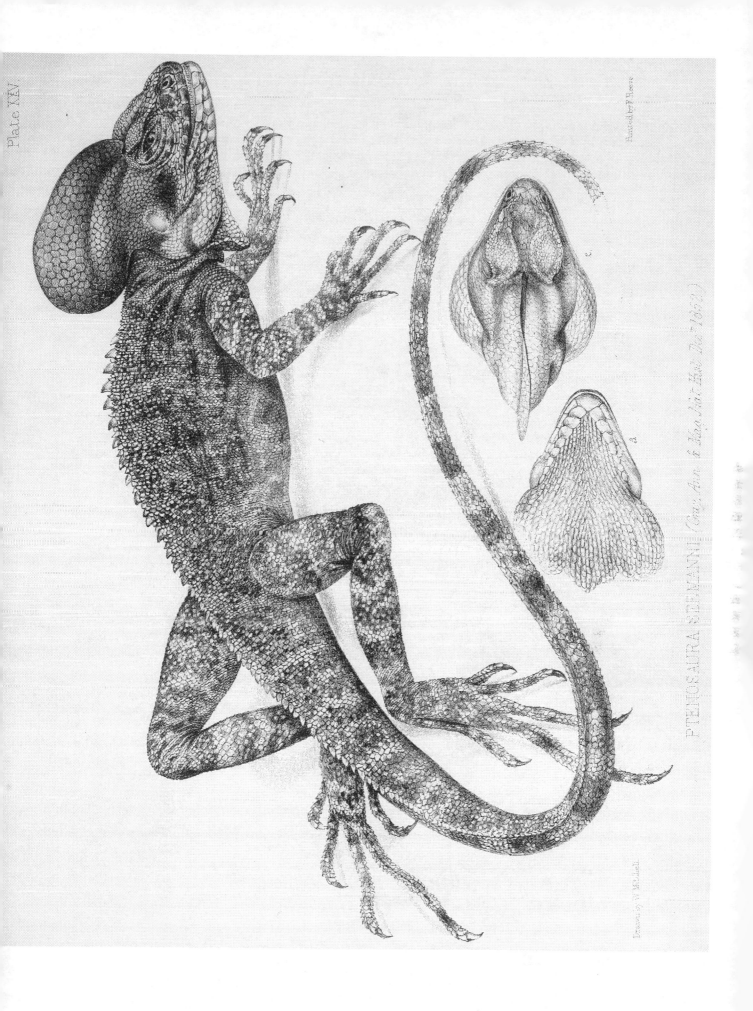

PTEROSAURA STEJNEGERI (*Cner. Ann. & Mag. Nat. Hist. Dec. 1852.*)

Plate XXVI

LOPHOSAURA COODRDCIL (Gray Ann & Mag. Nat. Hist. Dec.1852.)

Plate XXVII.

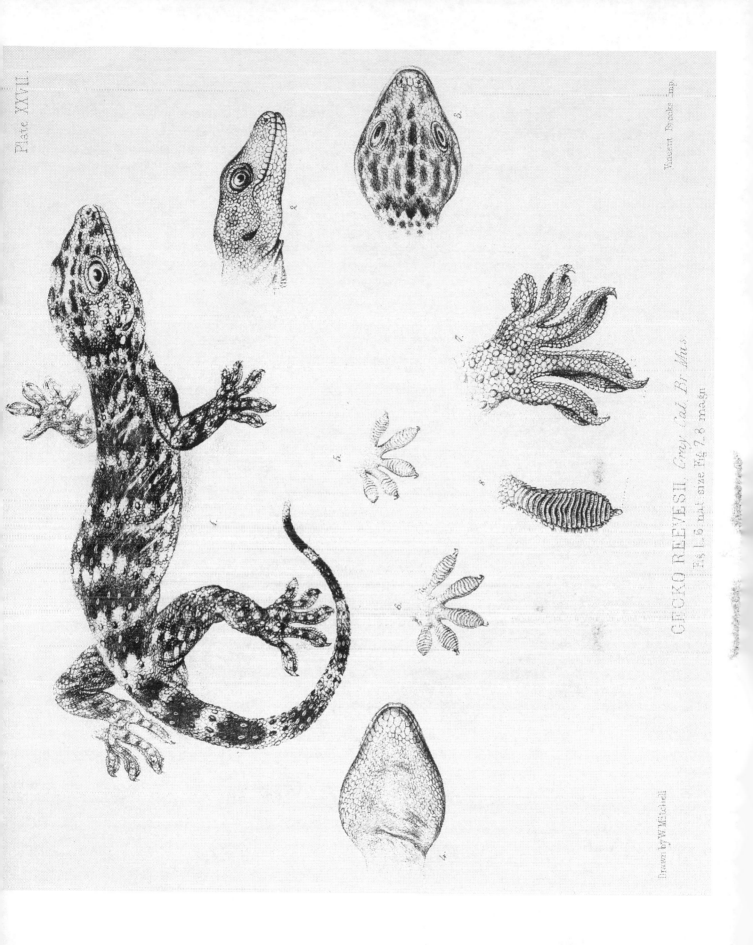

GECKO REEVESII. *Gray. Cat. Br. Mus.*
Fig 1-6 nat size. Fig 7.8 magn

ANHELLA PULCHRA Gray Fig 1 nat size Fig 2 5 magn
TETRODON VIRGATUS (Richardson) Fig 6 7 nat size Fig 8 mag

Pl. XXIX

ANCHISOMUS MULTISTRIATUS *(Kaup)*

nat. size

Drawn by W Mitchell

Vincent Brooks Imp

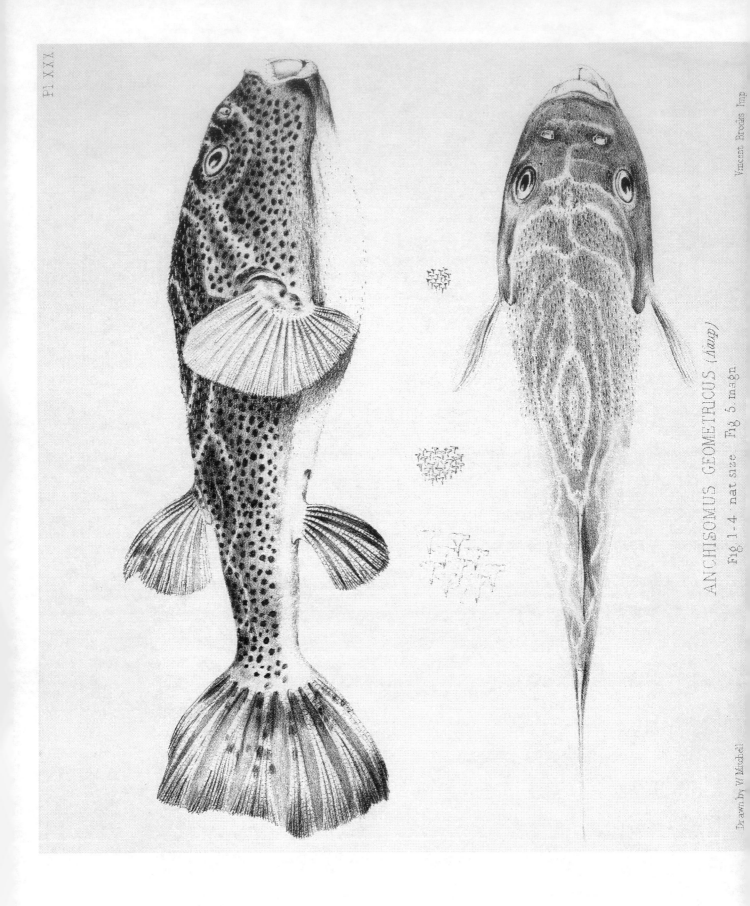

Pl XXX

Vincent Brooks Imp

ANCHISOMUS GEOMETRICUS (*hawp*)

Fig 1-4 nat size Fig 5 magn

Drawn by W Mitchell

Pl. XXXI.

Fig 1..3 ANCHISOMUS CAUDACINCTUS   Fig. 4..6 ANCHISOMUS RETICULARIS (*large*)

Drawn by W Mitchell   Vincent Brooks Imp

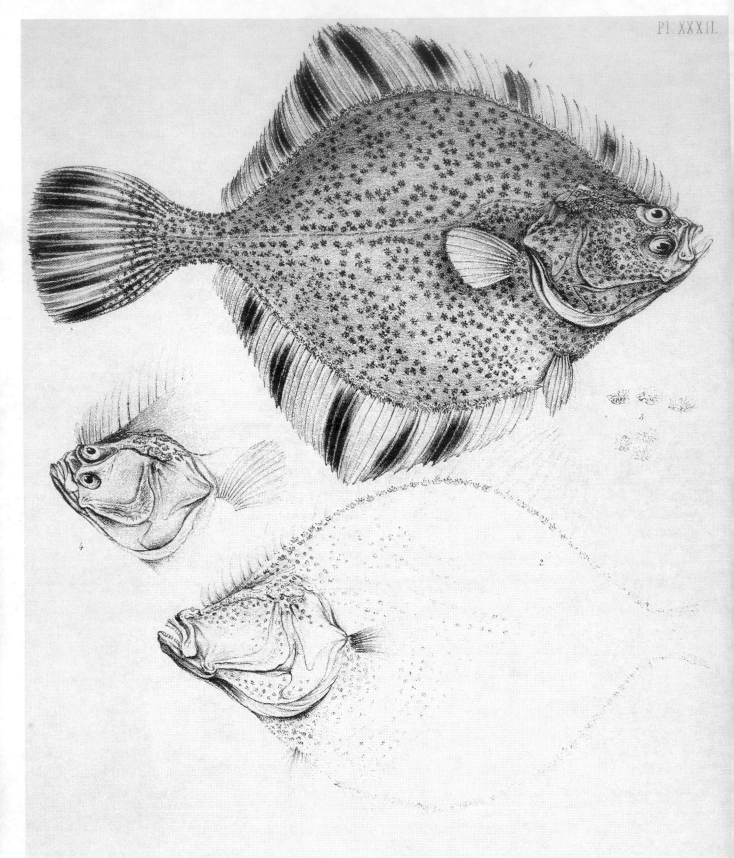

Pl. XXXII.

Fig 1. 3 PLATESSA STELLATA *(Pallas)*   Fig 4 PLATESSA GLACIALIS *(Pallas)*

Drawn by W Mitchell   Vincent Brooks Imp

Pl. XXXII

Printed in the United States
By Bookmasters